G

Experimental Phycology: a Laboratory Manual

EXPERIMENTAL PHYCOLOGY
A Laboratory Manual

CHRISTOPHER S. LOBBAN

University of New Brunswick, Saint John, Canada

DAVID J. CHAPMAN

University of California, Los Angeles, USA

BRUNO P. KREMER

Universität Köln, West Germany

Sponsored by the Phycological Society of America

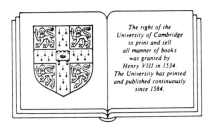

The right of the
University of Cambridge
to print and sell
all manner of books
was granted by
Henry VIII in 1534.
The University has printed
and published continuously
since 1584.

CAMBRIDGE UNIVERSITY PRESS

Cambridge

New York   New Rochelle   Melbourne   Sydney

Published by the Press Syndicate of the University of Cambridge
The Pitt Building, Trumpington Street, Cambridge CB2 1RP
35 East 57th Street, New York, NY 10022, United States of America
10 Stamford Road, Oakleigh, Melbourne 3166, Australia

First published 1988

Printed in the United States of America

Library of Congress Cataloging-in-Publication Data
Experimental phycology.
Includes index.
1. Algology—Laboratory manuals.   I. Lobban,
Christopher S.   II. Chapman, David J.   III. Kremer,
Bruno P.   IV. Physological Society of America.
[DNLM: 1. Algae—laboratory manuals.   QK 565.2 E96]
QK565.2.E97   1987   589.3′078   87-10286
8

British Library Cataloguing in Publication Data

Experimental phychology : a laboratory manual.
1. Algology—Laboratory manuals
I. Lobban, Christopher S.   II. Chapman,
David J.   III. Kremer, Bruno P.
589.3′028   QK565.2

ISBN 0-521-34333-X hard covers
ISBN 0-521-34834-X paperback

CONTENTS

LIST OF CONTRIBUTORS

Horst Bannwarth, Universität zu Köln, Institüt für Naturwissenschaften und ihre Didaktik, Abteilung für Biologie, 5000 Köln 41, Federal Republic of Germany (Exp. 31).

David J. Chapman, Dept. of Biology, University of California, Los Angeles, CA 90024, USA (Exps. 9, 10, 13, 16, 22).

Clinton J. Dawes, Dept. of Biology, University of South Florida, Tampa, FL 33620, USA (Exp. 7 App.).

Robert E. DeWreede, Dept. of Botany, University of British Columbia, Vancouver, BC V6T 2B1, Canada (Exp. 19).

Len V. Evans, Dept. of Plant Sciences, Baines Wing, The University, Leeds LS2 9JT, UK (Exps. 12, 14, 20).

David J. Garbary, Dept of Biology, St. Francis Xavier University, Antigonish, NS B2G 1C0, Canada (Exp. 19).

Paul J. Harrison, Dept. of Oceanography, University of British Columbia, Vancouver, BC V6T 2B1, Canada (Exp. 2, 3, 8, 22).

Eric C. Henry, Universität Konstanz, Fakultät für Biologie, Universitätsstrasse 10, Postfach 5560, D-7750 Konstanz, Federal Republic of Germany (Intro.).

Gunter O. Kirst, Fachbereich 2 (Biologie-Chemie) – Botanik, Universität Bremen, Leobenerstrasse NW 2, D-2800 Bremen, Federal Republic of Germany (Exps. 23, 24, 25, 26, 27).

J. N. Keen, Dept. of Plant Sciences, Baines Wing, The University, Leeds LS2 9JT, UK (Exp. 14).

Bruno P. Kremer, Universität zu Köln, Institüt für Naturwissenschaften und ihre Didaktik, Abteilung für Biologie, 5000 Köln 41, Federal Republic of Germany (Exp. 11, 15, 30).

Maureen A. Leupold, Ward's Natural Science Establishment Ltd., 5100 West Henrietta Road, P.O. Box 92912, Rochester, NY 14692-9012, USA (Exp. 5).

Christopher S. Lobban, Division of Sciences, University of New Brunswick, Saint John, NB E2L 4L5, Canada (Intro.).

Klaus Lüning, Biologische Anstalt Helgoland Zentrale, Notkestrasse 31, D-2000 Hamburg 52, Federal Republic of Germany (Exp. 28).

Geoffrey I. McFadden, Plant Cell Biology Centre, School of Botany, University of Melbourne, Parkville, 3052 Vic., Australia (Exp. 32).

Michael Melkonian, Westfälische Wilhelms-Universität, Botanisches Institüt, Schlossgarten 3, D-4400 Münster, Federal Republic of Germany (Exp. 32).

Dieter G. Müller, Universität Konstanz, Fakultät für Biologie, Universitätsstrasse 10, Postfach 5560, D-7750 Konstanz, Federal Republic of Germany (Exp. 29).

David J. Rawlence, Division of Sciences, University of New Brunswick, Saint John, NB E2L 4L5, Canada (Exp. 6).

Ingo B. Reize, Westfälische Wilhelms-Universität, Botanisches Institüt, Schlossgarten 3, D-4400 Münster, Federal Republic of Germany (Exp. 32).

Susan Schoen, Dept of Biological Sciences, University of California, Santa Barbara, CA 93106, USA (Exp. 1).

Nanette Sterman, Dept. of Biological Sciences, University of California, Santa Barbara, CA 93106, USA (Exp. 4).

Martin L. H. Thomas, Division of Sciences, University of New Brunswick, Saint John, NB E2L 4L5, Canada (Exp. 7).

Terry E. Thomas, Dept. of Oceanography, University of British Columbia, Vancouver, BC V6T 2B1, Canada (Exp. 3).

Herbert Vandermeulen, Huntsman Marine Laboratory, St. Andrew's, NB E0G 2X0, Canada (Exp. 17).

J. N. C. (Ian) Whyte, Fisheries and Oceans, Pacific Biological Station, Nanaimo, BC V9R 5K6, Canada (Exp. 18).

PREFACE

The experimental side of phycology has developed to such an extent over the last several decades that a need has arisen for a laboratory manual suitable for undergraduate teaching. In the view of the editors, such a manual would give step-by-step instructions for students and would specify all the preparations needed for specific experiments. The experiments would ideally be suitable for small campuses where equipment might be minimal as well as for more sophisticated laboratories; would teach the students a range of important laboratory techniques as well as illustrating key physiological or biochemical properties of algae; would use widely available or readily purchased marine and freshwater macroalgae and plankton; and would be tested to insure that all the necessary details had been included so that an instructor could expect the laboratory to work the first time.

We believe that most of the experiments satisfy most of these criteria. Most have come from teaching laboratories and most have been tested. They range from simple phytoplankton stains to elaborate studies of nucleocytoplasmic interactions. Thus, some can be quickly set up, whereas others require much preparation and effort. The most expensive equipment called for is a liquid scintillation spectrometer; the majority of the experiments, however, require only simple glassware, microscopes, centrifuges, and spectrophotometers.

Our procedure in assembling the manual was to solicit offers in the form of short drafts or direct student handouts, and to select those that best satisfied our criteria. We would like to take this opportunity to thank not only those teachers whose experiments were included, but also those whose experiments were turned down. The patience of the authors in revising their manuscripts toward a standard format is gratefully acknowledged.

Thanks are also due to colleagues who tested or reviewed the experiments: Frances Fulton, Kevin Gellenbeck, Aaron Gibor, Kathy Griffiths, Larry Liddle, Steven Manley, Anna Palmisano, Grant Steen, Jo-Anne Stevens, Carrie Theis and Robert Trench.

A NOTE ON SI

The Système International d'Unités (SI) prescribes certain base and supplementary units for use in scientific work. Some units that were excluded from SI remain useful, however, particularly in the practical aspects of laboratory work such as measuring volume in liters rather than cubic meters, measuring time in minutes and hours, and measuring lengths in centimeters. We have therefore retained such older metric units. Some conversions are:

$$1 \text{ L} = 1 \text{ dm}^3 = 1 \times 10^{-3} \text{ m}^3$$
$$1 \text{ mL} = 1 \times 10^{-6} \text{ m}^3 = 1 \text{ cm}^3$$
$$1 \text{ } \mu\text{L} = 1 \text{ mm}^3$$
$$1 \text{ } M = 1 \text{ mol dm}^{-3} = 1 \text{ kmol m}^{-3}$$
$$1 \text{ m}M = 1 \text{ mol m}^{-3}$$
$$1 \text{ g L}^{-1} = 1 \text{ kg m}^{-3}$$
$$1 \text{ } \mu\text{Ci} = 3 \times 10^4 \text{ Bq (30 kBq)}$$

For further information, see Chapter 2 in J. Gareth Morris (1974), *A Biologist's Physical Chemistry*, 2nd ed., Arnold, London; or Appendix A in F. B. Salisbury & C. W. Ross (1985), *Plant Physiology*, 3rd ed., Wadsworth, Belmont, CA.

WRITING A LABORATORY REPORT

Christopher S. Lobban

Division of Sciences, University of New Brunswick, Saint
John, NB, Canada, E2L 4L5

INTRODUCTION

A scientific study is not complete until it has been written up and, in the case of original research, published. In your class laboratory you are training to do original research and your laboratory reports are practice for writing papers and theses. Even if you have a sound grounding in English composition, you cannot expect to write good scientific reports at once because good style in science is different from good style in, say, short-story writing or literary criticism. Even though journals differ in minor details of manuscript presentation, there is a consensus on overall style, which has been embodied in books such as the *CBE Style Manual.* Of course, good science writing also requires a good command of language, but elegant prose is not essential. What is essential is clarity, and this is best achieved with short declarative sentences. Layers of meaning are inappropriate; unambiguous statements are required.

Clarity in writing also promotes clarity in thought. Writing your report brings into focus the strengths of your results—and also their shortcomings, which should be frankly admitted. Remember that science is a quest for understanding, and thus has no place for falsehood or deceit. Also keep in mind the human capacity for bias. When subjective judgment is involved in observation, we all have a subconscious tendency to err in favor of expectations. Even when numerical results are being recorded, clerical errors can occur and are likely to be biased – as anyone who checks the addition of sales clerks knows (Goldstein & Goldstein 1984).

The objective of this chapter is to present the basics of scientific style so that your laboratory reports can be practice papers. It is necessarily brief and you should take the time to read at least Day's (1983) entertain-

1

ing and informative manual, Chapters 6–14, 26, and 27, or Farr (1985), Chapters 1–4.

IN GENERAL

Your report should be organized into an Introduction, Materials and Methods, Results, and Discussion. You follow these sections with the References, Tables, and Figures. Full scientific papers always also have an abstract, but your report will be more like a "Note" or "Brief Communication," and probably does not need an abstract. (If an abstract is appropriate, you should compose it *last,* even though it is placed first in the paper.)

As with any writing, concise expression takes more time than verbosity. Always write a draft of your report, then edit it, tightening wherever possible, before you hand in the final copy. Thinking that your first draft is as good as you can make it is a mark of arrogance, not genius!

The passive voice is often used in scientific reports, but the active voice is more direct and is usually shorter. Learn to use the active voice. "I found" is better than the vague phrase "it was found that." Verb tense in scientific writing is, as Day (1983, Ch. 26) points out, tricky. What you did and what you found should always be in the past tense. If you are citing statements from the published literature, as in, "*Ulva* contains chlorophyll *b,*" use the present tense. When you attribute the finding, as in, "Brown (1977) *found* that *Ulva* contains chlorophyll *b,*" use the past tense. You also use the present tense when explaining what your tables and figures show.

Cite literature by author and date (as in the previous paragraph) or number your references in alphabetical order and cite them by number. (The former method is easier.) Do not use footnotes to cite literature in scientific writing. You do not need to cite the page number on which you found the information you are using, unless you quote verbatim. Look at citations in this manual to see how to use them smoothly. For papers with three or more authors, you may write the name of the first author, followed by et al. (e.g., Brown et al. 1986).

THE REPORT
Introduction

The purpose of the introduction in a scientific paper is to give the reader some background and context for your study. You should answer these two questions: Why was this study of interest? What is known to date? Most readers of a scientific paper are not experts in its subject. Depending on the journal, the introduction of a paper might have to be written for a wider or narrower audience, and you would slant it accordingly. For your lab reports, you should show that *you* understand the context of the lab.

Start with the broad context and quickly narrow it down to how your experiment fits in. Start positive! Give the significance of the kind of study you have done. Give references for facts you get from the literature. Do *not* try to cite every paper on the subject, but do cite the key pertinent works.

State the objective(s) of your experiment. You may indicate the outcome and you should study published papers to see how to do this without duplicating what will also be in the results. For example, if you were doing Experiment 12, you might write, "Pigments were extracted from algae grown under high and low irradiances to show the changes in pigment concentrations."

Be careful not to give methods, results, or discussion in the introduction.

Methods (or materials and methods)

For a lab report you may simply cite the chapter of the lab manual. You might use only part of a given experiment, or there might be differences from the instructions in the book. Therefore use a format like that used in published papers: "Pigments were extracted according to the instructions in Sterman (1987), except that. . . ." Record in your notes *every* difference from the instructions, but do not list them all in your report. Do give (in the Discussion) any difference that might have had a significant effect on the results. The principle behind writing methods is that someone reading your report should be able to repeat your work. (In theory, anyone should be able to follow your work through, but in practice no journal would publish the amount of detail needed in a lab manual to guide novices.)

Results

Do not simply supply your tables, figures and calculations. Start with a summary statement of your most significant result, citing the supporting data in parentheses. For instance, "Plants grown under dim light showed a dramatic increase in pigments compared to those in bright light (Table

1)." Then give other highlights of your results in the same way. Again, never repeat data in the text and in the tables or figures.

Tables and figures should be collected at the end of the report. Each should be numbered and should contain a short explanatory legend (caption).

Day (1983) gives some excellent pointers on when to use a table or a figure as opposed to making a statement in the text. The principle is, if you can say it all in a sentence, do so, and don't waste time and space constructing a table or graph.

Refer only to your own (or the class) results; do not give comparative data from the literature (the place for that is the Discussion). Occasionally, combining the Results and the Discussion may be appropriate but usually it is not and you should write them separately.

Discussion

The purpose of this section is to draw your conclusions from the results. Compare your data with published information and discuss discrepancies. If you did not find what you expected, consider why not. Evaluate your results frankly in the light of errors, omissions, and assumptions you made: how reliable are your results? Do not merely list sources of error, but estimate their potential impact: some may be trivial, others may be crucial. Draw whatever conclusions the data warrant, and do not make indefensible claims.

Start with a positive statement about your most significant result. The first sentence of the Discussion (together with the abstract, when there is one) is the most critical part of the report. It should tell the hasty reader the essence of what you found, in relation to what was already known. Never start by lamenting what went wrong, even in your humble lab report. You must say what went wrong, but weave it in later. If you have more than one kind of result (e.g., pigment content and growth rate) you can best organize your text by discussing the strengths and shortcomings of each in turn.

End by drawing your results back into the larger scientific context. Often the concluding paragraph will mention some far-reaching implication or critical further work that is needed to answer questions your study has raised. In part, you return to the starting point. Never leave your text hanging at the end of the list of shortcomings. Leave the reader with a positive memory of your work.

References

The references are an important part of any paper and must be given fully and checked carefully. Wasting time in a library trying to track down an incorrect reference is very annoying.

The precise style of the citations varies from journal to journal (e.g., whether there are periods or parentheses around the date, whether second author's initials follow or precede the surname). We suggest you follow the style of *Journal of Phycology*. Whatever style you pick, be consistent.

References include only papers cited in the text (a bibliography, which is not appropriate in a scientific paper, may include works not cited). List them in alphabetical order of first author. If there are two or more papers by the same author, arrange them in chronological order. If that author has also coauthored papers, these should follow his/her most recent solo paper. For instance:

Williams, A.B. 1952. . . .
Williams, A.B. 1965. . . .
Williams, A.B. & C.D. White. 1946. . . .
Williams, P.G. 1981. . . .

Information must include all authors' names (do not use et al. in the references unless there are ten or so authors), the date of the publication, the title, and, if it is not a book, the book or journal in which it appears. Here are some fictitious examples. Note that second and subsequent lines of each reference are indented.

Brown, G. A. & R. B. Smith (1983). The effect of copper on algae. *Mar. Biol.* 22, 365–377.
– (1983). The effect of copper on algae. In, A. Jones (ed.) *Biological Effects of Copper*. Academic Press, New York, pp. 365–377.
– (1983). *The Effect of Copper on Algae*. Academic Press, New York.

CONCLUSION

Careful and thorough work in the laboratory deserves a careful and thorough report, neatly presented in the appropriate style. The instructions in this chapter should be only a part of your training in scientific writing, just as the laboratory work is only part of your technical training. Read the pertinent chapters of Day's (1983) book and, most of all, look at the style in published papers as you read them.

REFERENCES

CBE Style Manual Committee (1978). *Council of Biology Editors Style Manual: a guide for authors, editors, and publishers in the biological sciences*, 4th ed. Council of Biology Editors, Inc., Washington, D.C.

Day, R. A. (1983). *How to Write and Publish a Scientific Paper*, 2nd ed. ISI Press, Philadelphia, PA.

Farr, A. D. (1985). *Science Writing for Beginners*. Blackwell Scientific Publ., Oxford.

Goldstein, M. & I. Goldstein (1984). *The Experience of Science. An Interdisciplinary Approach*. Plenum Press, New York.

ALGAL CULTURES

Eric C. Henry

Fakultät für Biologie, Universität Konstanz, Postfach 5560,
D-7750 Konstanz, Federal Republic of Germany

INTRODUCTION

If algae are to be used as experimental organisms, they will often have to be maintained in the laboratory in good physiological condition for extended periods. This will require that the experimenter become competent in at least a few algal culture techniques. Detailed accounts of many isolation and culture techniques have been presented in the *Handbook of Phycological Methods* (Stein 1973). This introductory chapter therefore provides, for instructors and students: (1) specific procedures for producing quantities of microalgae that will be required in some of the experiments that follow; (2) some alternative procedures and improvements in techniques, which have been devised since the *Handbook* was written.

CULTURE OF PHYTOPLANKTON

Large quantities of phytoplankton can be easily grown from small unialgal or axenic starting inocula. Unialgal stocks can be obtained from various suppliers such as the University of Texas at Austin, TX, and Carolina Biological Supply Co., Burlington, NC.

Materials

Glassware. Borosilicate (e.g., Pyrex, Kimax) glassware is preferred by most workers, although ordinary flint glass and plastics are often satisfactory and may be necessary for large-volume cultures. Culture tubes and bottles

7

or Erlenmeyer flasks are used for smaller volumes, jars and carboys for larger volumes.

Temperature control. Temperature should be controlled to within a few degrees Celsius in order to obtain consistent results. A controlled-temperature chamber or room is best, but an air-conditioned laboratory will often suffice. The optimum temperature for a given species is the ideal for culture, but a great many species have been found to grow well at temperatures ranging from from 15°–25°C.

Light. Fluorescent lighting is usually used, although in the Northern Hemisphere a north-facing window often works well. A light intensity of about 60–80 μE m^{-2} s^{-1} (3000–4000 lx) is often used and will usually be obtained within 20–30 cm of standard daylight or cool-white fluorescent tubes. It should be noted that the light output of fluorescent tubes declines with age, but cultures are probably given excess light in many cases. Some species benefit from a dark period each day, so a photoregime of 18 h light:6 h dark is often used.

Mixing. As the density of cultures increases, agitation of the medium is usually advantageous. Shakers, magnetic stirrers or aeration can be used. Air supplies must be free of oil vapors from pumps.

Media. No single medium can be expected to support good growth of all freshwater or marine species. Ideally one should use a medium that has been found to support good growth of the species to be cultured. Often the supplier will specify the medium used. Otherwise, one can use a medium that is known to be suitable for a large number of species, such as Chu No. 10 (freshwater) or Erdschreiber (marine). (Additional information follows in Table 1.)

Water. Glass-distilled and deionized water should be used in preparing freshwater media, salt solutions, and soil extract. Seawater media should be made with water obtained from an unpolluted area, preferably offshore. Seawater should be filtered to remove particulates, using Number 1 filter paper or, preferably, a 10 μm filter.

Procedures

Sterile techniques. Bacteriological-quality sterile techniques must be used for axenic (bacteria-free) culturing, and they are usually required for unial-

Table 1. *Sample culture media.*

Chu No. 10 (freshwater) (Chu 1942)

Salt stocks[1]	g L^{-1}
Ca (NO$_3$)$_2$	0.04
K$_2$HPO$_4$	0.01 or 0.005
MgSO$_4$·7H$_2$O	0.025
Na$_2$CO$_3$	0.02
Na$_2$SiO$_3$	0.025
FeCl$_3$	0.0008

Erdschreiber Medium (marine) (After Starr 1964)

Salt stocks[1]	g L^{-1}
NaNO$_3$	0.2
Na$_2$HPO$_4$·12H$_2$O	0.03
Soil water[2]	50 mL L^{-1}

[1] Stock salt solutions are prepared for use at 1:1000 dilution (1 mL stock/L medium). Autoclave stock solutions and store in refrigerator in tightly capped vials. Autoclave water and add salt solutions aseptically when cool.

[2] To prepare soil water: Place 1 cm of garden or greenhouse soil, which has not been treated with chemical fertilizers or pesticides, in the bottom of a 1-L flask. Cover with 1 L water and autoclave for 30 minutes. Let stand for one day and decant supernatant soil water.

gal culture of freshwater species, as contaminating algae are often airborne. Less exacting techniques may suffice for marine species, especially if culturing takes place away from the ocean.

All glassware and implements should be autoclaved at 121° C for 20 min or baked in an oven at 160° C for 2 h, except for disposable plasticware supplied sterilized by the manufacturer. Metallic implements may be flame-sterilized between repeated uses. A sterile-transfer hood, which enables cultures to be handled in a steam of filtered air, may be required if introduction of contaminants is a recurrent problem.

Inoculation.

1. Agitate the stock culture to ensure uniform distribution of cells to each sample.
2. If culturing axenically, remove the cap from the stock culture vessel and flame the opening briefly.

3. Using a sterile pipette, remove a measured aliquot of the stock culture. Flame the opening of the stock vessel and cap aseptically.
4. Squirt the contents of the pipette into a vessel containing sterile medium, again flaming the opening briefly.
5. Cap the culture vessel aseptically and place in culture facility.

SUPPLEMENTARY NOTES ON CULTURE OF ALGAE

Equipment

Microscopes and accessories. Significant improvement in equipment available for algal culture has resulted from the recent widespread interest in animal tissue culture. The use of inverted microscopes, in combination with plastic culture vessels of various designs, makes many observation and manipulation techniques much easier, faster, and more effective.

Tissue-culture inverted microscopes are equipped with long working-distance condensers and objectives (up to 40× magnification) so that *in situ* observations can be made on cultures growing in plastic (usually polystyrene) vessels, or through the bottom of glass microscope slides carrying preparations undergoing isolation or other manipulations. Such *in situ* observation avoids contamination and mechanical disturbance of the preparation, which can result from the use of water-immersion objectives on an upright microscope. Manipulations are further facilitated by the erect image formed by the inverted microscope, in contrast to the inverted image of the conventional upright microscope.

Several manufacturers (e.g., Nikon, Olympus, Zeiss, Swift) now offer simple and fairly inexpensive inverted microscopes suitable for student use; most may be equipped with phase–contrast optics and photomicrography attachments. Research–quality models may be equipped with any of the usual accessories such as differential interference optics, fluorescence illuminators and filters, and flash illuminators.

Plastic (polystyrene) culture vessels are superior to glass for use with inverted microscopes in most situations, because optically flat surfaces are required. Petri dishes of various sizes can be used to observe cultures growing in liquid media or agar media poured in thin layers. Flat-sided tubes (e.g., Lux Ambitube) may also be used in this way. For larger volumes (greater than about 20–30 mL), large petri dishes (e.g., Lab Tek, 100 × 25 mm, 80–100-mL capacity) and flat–sided tissue culture flasks with a capacity of 75–500 mL are available. Very small volumes (as little as 0.01 mL) and large numbers of replicates or isolates are easily handled with multi-well test plates or dishes.

Stereomicroscopes remain useful for initial isolation procedures of larger forms. High-intensity quartz-halogen illuminators, often with fiber-optic light guides are now available, helping to avoid overheating specimens.

Culture chambers. Culture chambers with electronic controls, which are more accurate and reliable than mechanical controls, have recently become available. The use of several small reach-in chambers, instead of a few walk-in chambers, makes it possible to keep replicates of valuable cultures in separate places and facilitates setting up a variety of temperatures and photoperiods for experimental work.

Extreme (up to 5°C) day/night temperature fluctuations may occur in some inexpensive chambers equipped with small air–circulating fans and fluorescent light fixtures with internal ballasts. This problem can be largely remedied by careful repositioning or removal of some air baffles to improve circulation, by moving the temperature sensors so that they can respond more quickly to temperature changes in the airstream, and by moving the ballasts outside the chamber.

For microscopy under controlled-temperature conditions, a less expensive (and often more comfortable) alternative to a walk-in chamber is a controlled-temperature stage for the microscope. Such a stage can be readily made using a thermo–electric cooling/heating module (e.g., Bailey Instruments TS-2). Fogging of the image from water condensation on cold culture vessels can be prevented by wiping the surface with a tissue wet slightly with full-strength Kodak Photoflo wetting agent.

Isolation

Culture of algae in enriched media for periods of more than a few days requires unialgal cultures. These will generally suffice for most experimental work unless nutritional physiology or growth regulation is to be studied. It is much more difficult to achieve and maintain axenic conditions. Moreover, some algae have been found to develop abnormally in the absence of bacteria (Provasoli & Pintner 1980).

Benthic forms. Isolations are preferably of spores or other reproductive stages that can be isolated as unicells. The hanging-drop technique (Wynne 1969) is a simple and effective method for obtaining contaminant-free isolates: Microscope coverslips are placed in petri dishes, each coverslip stuck to the bottom of the dish by the surface tension of a droplet of water on the underside. A drop of medium containing a small piece of fertile thallus, or a suspension of swarmers, is placed on each coverslip and the dishes are inverted carefully so that the drops remain on the coverslips. A piece of damp paper toweling or filter paper placed on the lid of the inverted dish prevents evaporation of the drops. Swarmers will settle on the coverslips, usually around the margin of the drop, while nonmotile contaminants (diatoms, etc.) sink to the bottom of the suspended drop and planktonic forms remain in suspension. The drops can be examined under a stereomicroscope or the low power objective of an upright compound microscope, or they can be briefly inverted and examined with the 40× objective of an inverted microscope. When the swarmers have settled, the water drop, with contaminants, can be removed by a flick of the coverslip or by gently washing it in sterile medium. Individual swarmers can be isolated by breaking the coverslips into small fragments.

An alternative method is to allow swarmers to settle on coverslip fragments strewn across the bottom of a petri dish. Fragments bearing only the desired species can be found with the inverted microscope and removed with fine forceps into separate culture vessels.

Isolates can also be made from multicellular forms that have diffuse or terminal meristems. The youngest portions of a thallus usually are the most vigorous and free of contaminants. Appropriate pieces are examined and cleaned with a soft brush or a stream of sterile medium, or they may be shaken vigorously in a culture tube half-filled with medium. Even more effective is the use of a small vial (23 × 13 mm) that fits snugly when placed horizontally in the rubber cup of a laboratory tube mixer (e.g., Vortex jr.). When shaken for a minute or two, thallus pieces are cleaned more effectively and are damaged much less than by other methods, including sonication.

Planktonic forms. Many planktonic species can be isolated on agar by conventional plating or streaking techniques (Hoshaw & Rosowski 1973). Species that cannot be grown on agar must be isolated by other methods. Individual cells or colonies that are large enough to be seen with a stereomicroscope can be drawn into fine pipettes manipulated by hand and isolated by sequential transfers into drops of sterile medium (Hoshaw & Rosowski 1973). Smaller forms are more easily isolated with a fine pipette mounted on the condenser of an inverted microscope, so that its tip is on the optical axis (Throndsen 1973). By adjusting the height of the condenser, the tip is brought into focus in a drop of medium containing the cells to be isolated, and the target cells are moved to the pipette by using the mechanical stage of the microscope as a micromanipulator, adjusting the position of the drop of medium that is placed on a microscope slide or in a petri dish.

Species of phytoplankton can be separated in larger quantities by centrifuging in a density gradient formed in a modified colloidal silica sol such as Percoll (Whitelaw et al. 1983).

Some contaminants can be eliminated by chemical means. Cyanophyceae are vulnerable to large dosages of antibiotics (Page 1973). Diatom cell division is arrested by germanium dioxide (Lewin 1966). It should be noted, however, that GeO_2 exists in two different crystal forms, one of which is practically insoluble (Weast 1977), and crystals often precipitate from stock solutions so that it is difficult to control the concentration in the medium. Fortunately, effective concentrations can be easily achieved by placing a few crystals in each culture dish or tube from which diatoms are to be eliminated. Upon death of the diatoms within a few days, a change of medium removes sufficient GeO_2 that brown algae, which may be severely inhibited by GeO_2 (Markham & Hagmeier 1982), grow well.

If diatoms are to be cultured, sodium metasilicate is added to the medium. If the stock solution is acidified as recommended by McLachlan (1973), it polymerizes, and only slowly (over about 2 days) depolymerizes in seawater (Suttle et al. 1986).

Tris (*tris*-[hydroxymethyl] aminoethane) is a frequently used buffer in artificial media (e.g., McLachlan 1973), but it has some undesirable properties, including toxicity to some algae, poor buffering capacity below pH 7.5, and competitive inhibition of K^+ uptake (McFadden & Melkonian 1986). A more suitable buffer, as demonstrated by McFadden & Melkonian, is Hepes (*N*-2-hydroxyethylpiperazine-*N'*-2-ethanesulfonic acid).

Culture media

Freshwater. Numerous recipes for growth media have been published and several widely used formulations have been presented in convenient form by Nichols (1973) and Bold & Wynne (1978). Recent new developments include the use of new buffers (Smith & Foy 1974, Cassin 1974), and unenriched natural waters with frequent transfers to retain normal morphologies (Trainor & Shubert 1974).

Marine. Marine media are either seawater with nutrients and buffers added, or are formulated from mixtures of salts and distilled water. Seawater is intrinsically well-suited for culture and comes ready-mixed, but it is both complex and variable and sometimes may be contaminated, so workers in regions remote from the sea, or who are concerned with phys-iological and biochemical problems, often prefer to use synthetic media. Many of the synthetic media, and several of the media based on seawater, have ionic ratios significantly different from seawater because of added nutrient salts. Recipes for both enriched seawater and synthetic media have been conveniently presented by McLachlan (1973) and Bold & Wynne (1978). Additional synthetic media have been proposed for grow-ing coralline red algae (Woelkerling et al. 1983) and phytoplankton (Har-rison et al. 1980). Of particular interest is a synthetic medium for which the availability of each ion in solution, which may differ by orders of mag-nitude from the nominal concentration in the recipe, may be calculated (Morel et al. 1979, Kuwabara & North 1980).

Erdschreiber medium (Table 1), which consists merely of seawater with nitrate, phosphate, and soil-water added, will grow many species and it is very easy to prepare. Other factors besides the choice of a particular medium may be important for good growth. Experience has shown that growth of macrophytes may be affected by the density of the culture (this is frequently difficult to quantify with macrophytes), and many species show improved growth, or more normal morphology and better repro-ductive performance, in larger volumes of medium (up to several liters). Reduction of nutrient enrichment to as little as one-quarter the usual con-centration may also be beneficial, and larger specimens usually show enhanced growth and more normal morphology and texture in cultures that are stirred.

REFERENCES

Bold, H. C. & M. J. Wynne (1978). *Introduction to the Algae*. Prentice-Hall, Englewood Cliffs, N.J.

Cassin, P. (1974). Isolation, growth and physiology of acidophilic *Chlamydo-monas. J. Phycol.,* 10, 439–447.

Chu, S. P. (1942). The influence of the mineral composition of the medium on the growth of planktonic algae. I. Methods and culture media. *J. Ecol.,* 30, 284–325.

Harrison, P. J., R. E. Waters & F. J. R. Taylor (1980). A broad spectrum artificial seawater medium for coastal and open ocean phytoplankton. *J. Phycol.,* 16, 28–35.

Hoshaw, R. W. & J. R. Rosowski (1973). Methods for microscopic algae. In, J. R. Stein (ed.) *Handbook of Phycological Methods. Culture Methods and Growth Measurements.* Cambridge University Press, Cambridge, pp. 53–67.

Kuwabara, J. S. & W. J. North (1980). Culturing microscopic states of *Macrocys-*

tis pyrifera (Phaeophyta) in Aquil, a chemically defined medium. *J. Phycol.*, 16, 546–549.

Lewin, J. (1966). Silicon metabolism in diatoms. V. Germanium dioxide, a specific inhibitor of diatom growth. *Phycologia*, 6, 1–12.

Markham, J. W. & E. Hagmeier (1982). Observations on the effects of germanium dioxide on the growth of macro-algae and diatoms. *Phycologia*, 21, 125–130.

McFadden, G. I. & M. Melkonian (1986). Use of Hepes buffer for microalgal culture media and fixation for electron microscopy. *Phycologia* 25, 551–57.

McLachlan, J. (1973). Growth media—marine. In, J. R. Stein (ed.) *Handbook of Phycological Methods. Culture Methods and Growth Measurements*. Cambridge University Press, Cambridge, pp. 25–51.

Morel, F. M. M., J. G. Reuter, D. M. Anderson & R. Guillard (1979). Aquil: a chemically defined phytoplankton culturing medium for trace metal studies. *J. Phycol.*, 15, 135–141.

Nichols, H. W. (1973). Growth media—freshwater. In, J. R. Stein (ed.) *Handbook of Phycological Methods. Culture Methods and Growth Measurements*, Cambridge University Press, Cambridge, pp. 7–24.

Page, J. Z. (1973). Methods for coenocytic algae. In, J. R. Stein (ed.) *Handbook of Phycological Methods. Culture Methods and Growth Measurements*. Cambridge University Press, Cambridge, pp. 105–126.

Provasoli, L. & I. J. Pintner (1980). Bacteria induced polymorphism in an axenic laboratory strain of *Ulva lactuca* (Chlorophyceae). *J. Phycol.*, 16, 196–201.

Smith, R. V. & R. H. Foy (1974). Improved hydrogen ion buffering of media for the culture of freshwater algae. *Br. Phycol. J.*, 9, 239–245.

Starr, R. C. (1964). The culture collection of marine algae at Indiana University. *Amer. J. Bot.*, 51, 1013–44.

Stein, J. R. (Ed.) (1973). *Handbook of Phycological Methods. Culture Methods and Growth Measurements*. Cambridge University Press, Cambridge, xii+ 448 pp.

Suttle, C. A., N. M. Price, P. J. Harrison & P. A. Thompson (1986). Polymerization of silica in acidic solutions: a note of caution to phycologists. *J. Phycol.* 22, 234–237.

Throndsen, J. (1973). Special methods—micromanipulators. In, J. R. Stein (ed.) *Handbook of Phycological Methods. Culture Methods and Growth Measurements*. Cambridge University Press, Cambridge, pp. 139–144.

Trainor, F. R. & L. E. Shubert (1974). *Scenedesmus* morphogenesis. Colony control in dilute media. *J. Phycol.*, 10, 28–30.

Weast, R. C. (1973). *Handbook of Chemistry and Physics*. 54th ed. CRC Press, Cleveland.

Whitelaw, G. C., T. Lanaras & G. A. Codd (1983). Rapid separation of microalgae by density gradient centrifugation in Percoll. *Br. Phycol. J.*, 18, 23–28.

Woelkerling, W. J., K. G. Spencer & J. A. West (1983). Studies on selected Corallinaceae (Rhodophyta) and other algae in a defined culture medium. *J. Exp. Mar. Biol. Ecol.*, 67, 61–77.

Wynne, M. J. (1969). Life history and systematic studies of some Pacific North American Phaeophyceae (brown algae). *Univ. Calif. Pubs. Bot.*, 50, 1–88.

1

Cell counting

Susan Schoen

Department of Biological Sciences University of California,
Santa Barbara, CA 93106, USA

FOR THE STUDENT

Introduction

Cell counting is a valuable and necessary technique for the study of phytoplankton number in both culture and field work. Experiments done on phytoplankton in culture require knowledge of the age and health of the population. Cell density (determined by cell counts) plotted against time yields a growth curve for the population. From this growth curve, the investigator can determine the age of the population at the time of the experiment. With cell density data, the experimenter can relate physiological processes to a per-cell basis. While counting cells, one can assess the health of the population by monitoring the relative numbers of dividing, broken or dead cells. Cell counting is a much used technique in phytoplankton culture work.

To perform cell counts on cultured phytoplankton, you will use some type of counting chamber and the compound microscope. The procedure involves: placing a volume of your culture in a hemocytometer, or counting chamber; counting the number of cells that appear in several microscope fields, so that you can obtain the mean number of cells per known volume; and converting your replicated counts to number of cells per mL. Your instructor will assign you an appropriate chamber and instruct you in its use.

16

About your cultures

If you are to maintain your own cultures for the duration of the laboratory, it is important that you learn proper techniques for keeping your cultures alive and healthy (See Culture Techniques). You should note in particular the following points:

1. Avoid contamination of your stock culture by ambient bacteria and fungi by using sterile techniques. Be sure to flame the culture flask during transfer of cultures. When obtaining a culture sample, never stick a pipette into the culture flask. Instead, pour a small amount of culture (mix it first) into a beaker and draw your sample from this supply. Also, never place the stopper on the counter. Hold it in your hand, and be sure not to touch the part of the stopper that will be reinserted into the flask.

2. Your culture may have become too dense for its present flask. Ask your instructor how long the culture has been growing and whether or not it should be split. If need be, split a culture by adding an equal volume of fresh medium, swirling, and pouring half of it into a new sterile flask. Remember to flame both flasks.

Procedure

1. Pour a well-mixed sample into a small beaker. (Remember to use sterile technique!) Add one small drop of diluted Lugol's solution to kill the cells.
2. Determine which counting chamber is appropriate for your culture (Guillard 1978). If you are counting larger phytoplankton such as chain-forming diatoms or dinoflagellates, the Sedgwick-Rafter chamber may be appropriate. Denser cultures of smaller cells are best counted with a hemocytometer. Regardless of which slide you choose, use a Pasteur pipette to bubble the sample and mix it. Follow the instructions that are specific to your chamber.

Hemocytometer

1. The hemocytometer has two counting grids onto which you should place your sample. Discard the first drop in the pipette, as it contains few phytoplankton. Then apply one drop to each grid at the point shown by the arrow in Figure 1.1. Check to see if your hemocytometer has the Fuchs-Rosenthal or Improved Neubauer ruling. The Fuchs-Rosenthal grid consists of 16 squares, each of which is further divided into 16 squares. The larger squares measure 1 mm along a side; therefore, the entire grid is 16 mm^2. The hemocytometer chamber is 0.2 mm deep, thus the volume of sample that covers the grid of 16 large squares is 0.0032 mL. The Improved Neubauer grid has nine squares, each

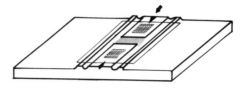

Fig. 1.1. Typical hemocytometer with cover slip. Arrow indicates point of application of phytoplankton sample.

1. mm along a side, and is further divided into 20 or 25 smaller squares. The chamber is 0.1 mm deep, therefore each grid (9 mm^2) holds 0.009 mL of sample (Guillard 1978).

2. Use low power to scan the grid in order to acquaint yourself with the setup and to identify the plankters that you will be counting. Make sure that you can distinguish cells from detritus. Decide if cells that fall on the line are "in" or "out." Are dividing cells counted as one cell or as two cells? What about dead cells? These are questions that you must answer for yourself. Whatever you decide, you must be consistent.

3. Use the magnification that enables you to count your cells most easily. Be careful not to crack the special hemocytometer cover slip. Always start with the ocular in the lowest position and move it upwards.

4. Use a counter to keep track of your results. Individual counts of at least 30 cells per unit area are statistically desirable. Counts above 30 cells per unit area may become unmanageable. Choose the unit area to be counted accordingly. Select areas to be counted randomly from both grids, and do at least 7–10 replicate counts. Calculate a mean, standard deviation and coefficient of variation from your counts. Your instructor or any basic statistics book can help you to understand the statistical implications of these measures. If you are working with a partner, check your results against his or hers. Make sure that you both counted the same unit area. If your values differ greatly from those of your partner, you might try again to improve your accuracy.

5. Multiply the mean of your counts by the factor which will give you an average number of cells in the entire grid. Multiply this value by 1 mL divided by the volume of the sample on the grid. Your calculations are:
   a) Mean number of cells/unit area × area of entire grid/unit area counted = number of cells/grid
   b) Number of cells/grid × mL/volume of sample on the grid = number of cells/mL

For example, with a Neubauer hemocytometer you might have counted cells in the 1 mm^2 squares. You would then multiply your mean count (let's say it was 20 cells) by nine to obtain the average number of cells over the entire grid, or 180 cells. The grid has a volume of 0.009 mL. Therefore, the number of cells per mL for your sample = 180 cells × (1 mL / 0.009 mL) = 2.0×10^4 cells mL^{-1}.

Sedgwick-Rafter slide

1. This counting chamber holds 1 mL of sample, has an area of 1000 mm^3, and is 1 mm deep (Fig. 1.2). To use it, pipette 1 mL of mixed sample into the chamber. Slide the special coverslip across the top of the chamber, making sure there are no bubbles trapped beneath the coverslip. Place the chamber on the microscope stage and let the phytoplankton settle to the bottom of the chamber.
2. This chamber has no grid lines; therefore, you must either count all the organisms in the chamber (if they are scarce) or use the area of the microscope field as a frame of reference. If a Whipple field is available for your microscope, you can use it instead of using the entire microscope field. Consult your instructor. If no Whipple field is available, you will determine the mean number of cells per microscope field, determine the area of the microscope field with a micrometer slide, and then convert your average counts per unit area to the number of cells per mL using the dimensions of the chamber.
3. As with the hemocytometer, you should aim for a number of counts that yields the most reliable statistics. Try for counts of approximately 30 cells. You need to count fields throughout the entire chamber. Imagine the chamber as a grid that is made up of four rows and four columns. Start in the upper left-hand corner and count four fields evenly spaced across the top row of the chamber. Return to the left-hand edge of the chamber and count another row beneath the first. Continue until you have counted the four rows, or a total of 16 microscope fields. Calculate a mean number of cells per microscope field.
4. Your instructor will help you measure the area of the appropriate microscope field using a micrometer slide. Calculate the area of the field.
5. You can now calculate the number of fields per area of the Sedgwick-Rafter chamber and thus the number of cells per mL.

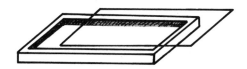

Fig. 1.2. Sedgwick-Rafter Chamber with coverslip.

For example, suppose you used a microscope field that had an area of 1 mm^2, and you counted an average of 15 cells per field. The Sedgwick-Rafter chamber has an area of 1000 mm^2 and holds 1.0 mL. Therefore, the number of cells per mL of sample is 15 cells $\times$ 1000 mm^2 $\div$ 1.0 mm^2 = 1.5 $\times$ 10^4 cells mL^{-1}.

Growth curves

It is likely that you will count cells from your culture during each laboratory period over the course of several weeks. Once you have obtained the data set of sequential cell counts, you can construct a growth curve from your data. You should plot the number of cells against time in days. From these curves, you can calculate two useful parameters of population growth: specific growth rate or growth constant (μ) and the division time or generation time (t_g).

Your plot of cell number against time probably shows an S-shaped growth curve. At first, the cells were not dividing. This is the "lag phase." Within a few days, the phytoplankton started to divide, and your curve probably shows a rapid increase in cell number over time. This is the exponential, or "log," phase and here the rate of increase in cell number is proportional to the number of cells present (which is increasing). Finally, your curve may show a plateau, or leveling off, of cell density. This final stage is the stationary phase and is indicative of a population that may have grown too dense for its container and that will be light or nutrient-limited.

The growth rate shown by your phytoplankton during the exponential phase is an important characteristic of your particular culture, as is the division time (or time for your cells to divide). You should calculate both the growth rate and division time for your culture using the following equations:

1. The growth rate can be calculated with the differential equation,

$$dX/dt = \mu X$$

   where X is the number of cells, μ is the growth rate and t is time in days. Rearrangement of this equation yields,

$$\mu = \ln X_2 - \ln X_1 / t_2 - t_1$$

   where X_2 and X_1 are cell densities at two times t_2 and t_1.
2. Division time, t_g, is the time for cells to divide (in days) and can be calculated from the growth rate (Guillard 1973):

$$t_g = 0.6931 / \mu$$

For example, you have two counts on day six and day eight (which coincide with the exponential phase of growth). The counts are 5.0×10^3 and 1.5×10^4 cells mL^{-1}, respectively. Your calculations for growth rate and division time are:

$$\mu = \ln X_8 - \ln X_6 / t_8 - t_6$$
$$= \ln 1.5 \times 10^4 - \ln 5.0 \times 10^3 / 2$$
$$= 1.0986 / 2$$
$$= 0.5493$$
$$t_g = 0.6931 / \mu$$
$$= 1.2618 \text{ days}$$

REFERENCES

Boney, A. D. (1975). *Phytoplankton*. Edward Arnold Publishers, London.

Guillard, R. R. L. (1973). Division rates. In, J.R. Stein (ed.) *Handbook of Phycological Methods*. Cambridge Univ. Press, Cambridge, pp. 289–312.

Guillard, R. R. L. (1978) Counting slides. In, A. Sournia (ed.) *Phytoplankton Manual*. UNESCO: Paris, pp. 182–189.

Vollenweider, R. A. (1969). *A Manual on Methods for Measuring Primary Productivity in Aquatic Environments*. Blackwell Scientific Publications, Oxford.

FOR THE INSTRUCTOR

Supplies

These are the materials necessary for each student or student pair.

1. phytoplankton culture (see Culture Techniques)
2. culture media and flasks
3. bunsen burner
4. 1–5 mL beaker (1)
5. hemocytometer or counting chamber
6. special cover slips for hemocytometer or Sedgewick Rafter chambers
7. Pasteur or graduated pipette and bulb
8. compound microscope
9. counter
10. micrometer slide
11. dilute Lugol's solution:
    Lugol's stock solution: dissolve 2.0 g potassium iodide and then 1.0 g iodine crystals in 20 mL distilled water.
    Dilute this stock to pale brown color, approximately two drops per mL.

Results and problems

The biggest problem is a culture that dies owing to contamination or neglect (i.e., it is not split in time). You should impress upon students the need to follow sterile techniques when working with their cultures. Cultures of different species will grow at different rates; for example, diatoms grow very rapidly, many dinoflagellates are slow growing (they may take weeks to reach the plateau phase of growth). You should be aware of the growth characteristics of your cultures if you expect to see a complete growth curve for the cultures. Also, there is flask-to-flask variability in growth rates of the same culture. Students should take the samples to be counted from the same culture each lab period. This requires that the volume of media be sufficient to provide about 10 mL per count for the duration of the experiment. Encourage the students to be sparing about the amount of culture they use each time.

Check to see that the students are careful and accurate in their counting. Be sure that they really are counting cells, not detritus. If they begin to see detritus in their cultures there may be contamination. At first this technique will take them at least half an hour, but, with practice, they will be able to do the counting very efficiently and can do it routinely at the start of a busy lab period. The more often they do the count, the better their growth curve will look. Don't pass up the opportunity to have them count each lab period. If possible they should count every day or two, especially for fast-growing algae.

This method is useful for all types of phytoplankton, though single cells are easiest to count. The hemocytometer is very helpful. If you need to use counting chambers other than the hemocytometer, consult Guillard (1978) and the instructions that come with the chamber.

2

Phytoplankton stains

P. J. Harrison

Oceanography Dept., University of British Columbia,
Vancouver, BC, Canada, V6T 2B1

FOR THE STUDENT

Introduction

Histological staining techniques provide a qualitative index of chemical composition (e.g., lipids, starch) and physiological state (e.g., cell viability) of phytoplankton. Although staining may not work for all species in natural assemblages, it is a highly effective method of estimating the viability of a few target species (e.g., large diatoms), which have previously been studied in the laboratory under defined environmental conditions.

Two commonly used stains for assessing cell viability are neutral red and Evan's blue (Crippen and Perrier 1974; Gallagher 1984). Neutral red is a vital (sublethal) stain that is taken up by living cells and that accumulates primarily in the vacuoles in 5–10 min. If the cells are dead, neutral red will stain the cytoplasm pink or red, thus distinguishing them from living cells. Evan's blue is a mortal stain that turns organic matter of dead cells deep blue. It is *excluded* by live cells, which possess a functional cell membrane.

Examples of stains that can be used to detect chemical constituents of phytoplankton are: iodine staining for starch and Nile blue A or Sudan IV for lipids. The starch will stain deep blue, while fat will stain red, and fatty acids, a bluish color.

The objective of this experiment is to determine cell viability and chemical constituents in phytoplankton.

Procedure

Cell Viability

1. You have been provided with a large diatom (e.g., *Ditylum brightwellii*), which has a prominent vacuole. Three cultures of this diatom have been provided: (1) logarithmically growing (all living cells); (2) senescent (many dead cells); and (3) heat killed cells (all dead cells).
2. Add 1 mL of 0.01% (w/v) neutral red stock solution to a 50 mL sample of culture to yield a final concentration of 2×10^{-6} g stain per mL seawater.
3. Add 1 mL of a 1% (w/v) stock solution of Evan's blue to a 20 mL sample of culture to yield a final concentration of 5×10^{-7} g stain per mL seawater.
4. Incubate samples for 10–20 min at room temperature, make a whole mount, and examine.

Chemical constituents

1. Add 1 or 2 drops of the KI/I_2 solution to 5 mL of the chlorophyte culture grown under high or low light. Wait a few minutes and then compare the amount of starch in the two cultures grown under the saturating and limiting irradiances. Make a wet mount and examine with a compound microscope.
2. Fix diatoms (logarithmically growing and nitrogen-starved cultures) in 2% formalin. Stain with Nile blue A for 20 minutes and then wash in water and mount in glycerol. Compare the amount of lipid in the nitrogen-starved culture and the logarithmically growing culture (control).

Questions

1. Did the neutral red and Evan's blue stains differentiate between living and dead cells? Discuss.
2. In the introduction, it was stated that viability stains may not work for all species. Explain this statement.
3. Did the nitrogen-starved culture have more lipid per cell than the nitrogen-sufficient culuture? Discuss.
4. Some cells may be in a physiological resting stage. Under which conditions do they enter this stage? What do you think the results of the viability tests would be for cells in this stage?

REFERENCES

Crippen, R. W. & J. L. Perrier (1974). The use of neutral red and Evan's blue for live-dead determination of marine phytoplankton. *Stain Technol.*, 49, 97–104.

Gaff, D. F. & O. Okong'O-Ogala (1970). The use of non-permeating pigments for testing the survival of cells. *J. Exp. Bot.,* 22, 756–758.

Gallagher, J. C. (1984). Patterns of cell viability in the diatom, *Skeletonema costatum* in batch culture and in natural populations. *Estuaries,* 7, 98–101.

McCully, M. E., L. J. Goff & P. C. Adshead (1980). Preparation of algae for light microscopy. In, E. Gantt (ed.) *Handbook of Phycological Methods. Developmental and Cytological Methods.* Cambridge University Press, New York, pp. 263–283.

O'Brien, T. P. & M. E. McCully (1981). *The Study of Plant Structure: Principles and Selected Methods.* Termacarphi Pty. Ltd, Melbourne.

Reynolds, A. E., G. B. Mackiernan & S. D. Van Valkenburg (1978). Vital and mortal staining of algae in the presence of chlorine-producing oxidants. *Estuaries,* 1, 192–196.

Shifrin, N. S. & S. W. Chisholm (1981). Phytoplankton lipids: interspecific differences and effects of nitrate, silicate and light-dark cycles. *J. Phycol.,* 17, 374–384.

Smith, B. A., M. L. Reider & J. S. Fletcher (1982). Relationships between vital staining and subculture growth during the senescence of plant tissue cultures. *Plant Physiol.,* 70, 1228–1230.

Widholm, J. M. (1972). The use of fluorescein diacetate and phenosafranine for determining viability of cultured plant cells. *Stain Technol.,* 47, 189–194.

NOTES FOR INSTRUCTORS

Materials

1. Microscope–with $40 \times$ objective
2. Microscope slides and cover slips
3. Pasteur pipettes
4. Graduated cylinders
5. Balance and weighing boats
6. Volumetric pipettes (0–5 mL)

Reagents and chemicals

1. 0.01% (w/v) neutral red in distilled water
2. 1% (w/v) Evan's blue in distilled water
3. 10 g KI
4. 5g I_2 } for starch test
5. 2% formalin (formalin = 40% formaldehyde)
6. 0.05% (w/v) Nile blue A
7. 1% (w/v) H_2SO_4 } for lipid test
8. glycerol (full strength)

Preparation of solutions

1. To make the KI/I_2 solution for the starch test, dissolve 10 g KI in 20

mL distilled water and add 5 g I_2. When these are dissolved, add 80 mL distilled water.

2. To prepare the Nile blue A solution for the lipid test, dissolve 0.05% w/v of Nile blue A in 1% w/v of H_2SO_4.

Cultures

1. See Culture Techniques, (pp 7–15) and Cell Counting (Exp. 1) for methods to determine logarithmic and senescent phases. Only a few mL of culture per student are required.

2. Choose any large diatom (because they have a large vacuole–e.g., *Ditylum brightwellii*) and grow it in two batch cultures. One culture should be in logarithmic phase and the other in senescent phase. Take part of the log phase culture and heat it at 50°C for 5 min (to produce heat-killed cells).

3. Choose any Chlorophyceae (e.g., *Brachiomonas* spp. because it is relatively large in size) and grow one culture under logarithmic growth conditions (>150 μE m^{-2} s^{-1}) so that it will contain maximum amounts of starch, and grow the other one under light limitation ($\sim$ 30 μE m^{-2} s^{-1}) so it will have a minimum of starch. Make sure both cultures are in log phase growth when the students examine them.

4. Choose any large diatom and grow it under logarithmic growth phase conditions, and grow another culture until it becomes nitrogen-starved–about 72 h of starvation after nitrogen in the medium reaches zero–in order to increase the amount of lipid per cell. Note: Allow one to two weeks to produce a nitrogen-starved culture, depending on its growth rate and the concentration of nutrients in the medium. See Experiments 1 and 4 to determine when cell growth has stopped, using *in vivo* fluorescence.

5. These stains have not been tried for macrophytes. The multilayered aspect of most thalli makes microscopic examination difficult.

3

Biomass measurements: protein determination

Paul J. Harrison & Terry E. Thomas

Oceanography Dept., University of British Columbia, Vancouver, BC, Canada, V6T 2B1

FOR THE STUDENT

Introduction

Certain biochemical constituents of algae can be used as an estimate of biomass. Chlorophyll *a* and protein are commonly used. Protein determinations are especially useful when working with physiological processes that are dependent on the activity of free protein in the cell. For example, enzyme activities are usually normalized to soluble protein, rather than chl *a*, cell numbers, or wet weight.

There are several different methods for measuring protein concentrations (Bergmeyer & Grabl 1983). Proteins have a number of basic properties in common, but they vary greatly in their amino acid composition and sequence, as well as in their size and shape. Since the different assay methods depend on different properties of the proteins, consistency between the methods may be low. The nitrogen content of proteins is relatively invariable at 0.16 g per g of protein. Thus, nitrogen determination by the methods of Kjeldahl or Dumas is an accurate, but rather complicated, procedure to determine protein concentration. The most common techniques are the spectrophotometric methods and measurements according to Lowry *et al.* (1951) or Bradford (1976). In the Lowry method, proteins react with copper ions in alkali through the reduction of the Folin-Ciocalteu phenol reagent (phosphomolybdic-phosphotungstic acid) by tyrosine and tryptophan residues (see Peterson, 1979, for a detailed review of the Lowry method). In the Bradford method, the protein is stained with a dye, Coomassie Brilliant Blue. This method is based on the shift of the absorption maximum of the dye from 465 to 595 nm, which occurs when the dye binds to the protein.

The basic assumptions underlying these techniques are:

1. Bovine serum albumin (BSA), the standard protein used to calibrate the method, has assayable properties equivalent to soluble algal protein. This is not true, since a single protein is often not representative of a mixture of proteins.
2. Marine algae do not contain compounds that interfere with the method. This also is not true.

Although the Bradford method stands out for its simplicity (one reagent) and greater sensitivity (about four times more than the Lowry method), it has some important limitations (Bergmeyer & Grabl 1983). The standard curves are nonlinear for many proteins, especially with more than 60 μg of protein. This inherent nonlinearity is caused by the reagent itself, as there is an overlap in the spectrum of the two different color forms of the dye, so that the background value for the reagent is continually decreasing as more dye is bound to protein. Absorbance may also vary with the age of the dye reagent.

A very serious problem with all assays based on dye binding is variation of response with different proteins, which can be severe. When studied with 23 different proteins, the standard deviation in estimates of protein concentration by the Bradford method was twice the value obtained by the Lowry method. Therefore, it is important to use a standard protein that gives a similar color yield as the protein or mixture under study. Unfortunately, bovine serum albumin is a poor standard in the Bradford method, since it gives a much higher color yield than most other proteins. (Note: it will be used in this lab for reasons of simplicity.) Bovine gamma globulin was recommended as a better standard, giving a more normal response and a linear calibration curve (Bio-Rad 1979).

The Lowry method is known to give a nonlinear calibration curve. Thus, for better precision it is advisable to assay at least three different dilutions of the sample. The nonlinear behavior appears to be inherent in the reaction mechanism (Peterson 1979). The color given by a protein in the Lowry assay depends on its content of tyrosine and tryptophan residues. Therefore, variation of response with various proteins can be quite significant. A large number of substances interfere with protein quantitation by the Lowry assay.

The objective of this experiment is to measure the protein content of tissue extracts of an alga (e.g., *Porphyra perforata*) using the Lowry and Bradford protein methods. The efficiency of the two methods will be compared.

Procedure

Extraction

1. Place an aliquot of known fresh wt of tissue (approximately 2 g blotted dry) and 1 g of sand in a clean mortar.
2. Add 5 mL of 0.3 M phosphate buffer (pH 6.8–7.0) and grind until a fine tissue-water slurry is produced. Add 20 mL distilled water. Grind again briefly.
3. Pour tissue extract into two 15 mL centrifuge tubes.
4. Centrifuge at high speed ($>3,000$ g) for 5 min (or until clear. A high speed centrifuge [$\sim 30,000 \times$ g] may be necessary). Decant supernatant (tissue extract).

Lowry method

1. Dilute tissue extract 1:1 with distilled water.
2. Set up five test tubes with 1 mL reagent C. Label two treatment and three blank tubes.
3. Add 0.2 mL extract to the treatment tubes and 0.2 mL distilled water to the blanks and mix, preferably with vortex mixer.
4. Let stand 10 min.
5. Add and mix 0.1 mL reagent D.
6. Let stand 0.5 h.
7. Read at 750 nm against water (blank).

Bradford method

1. Dilute tissue extract 1:1 with distilled water.
2. Set up four test tubes with 5 mL Coomassie Blue reagent. Label two treatment and two blank tubes.
3. Add 0.1 mL tissue extract to the treatment tubes and 0.1 mL distilled water to the blanks and mix, preferably with a vortex mixer.
4. Let stand 0.5 h.
5. Read absorbance at 595 nm against water (blank).

Standard curves. Take 0.5 mL of 100 g L^{-1} of bovine serum albumin stock and dilute to 100 mL. This equals 500 mg L^{-1}. Set up duplicate series of diluted standard solutions as follows:

Albumin standard	Water	Standard solutions
10 mL	0 mL	500 mg L^{-1}
5.0	5.0	250
2.5	7.5	125
1.0	9.0	50
0.5	9.5	25
0.1	9.9	5
0	10.0	0 (blank)

Treat each standard solution as if it were the diluted tissue extract. Analyze using the Lowry and Bradford Methods, Step 2 and on.

Report

1. Plot standard curves for both the Lowry and Bradford methods (absorbance versus mg protein L^{-1}).
2. Is one method more sensitive than the other? Is one more variable?
3. Calculate the free protein content of the *Porphyra perforata* thallus in mg protein g wet wt^{-1} for both methods. Protein may be expressed as mg protein g dry wt^{-1} or mg protein surface area^{-1}, if these parameters are measured.
4. Are the values calculated in Number 3 consistent with the differences found in the standard curves for the two methods? If not, suggest a possible reason.
5. Discuss environmental and physiological factors that could affect the protein content of marine algae or the thoroughness of extraction.

REFERENCES

Bergmeyer, J. & M. Grabl (eds.) (1983). *Methods in Enzymatic Analysis* (3rd edition) Vol. I. *Fundamentals*. Verlag Chemie, Weinheim, pp. 84–99.

Bio-Rad (1979). Bio-Rad Protein Assay. *Bio-Rad Bull.* 1069 EG, Bio-Rad Laboratories, Richmond, California.

Bradford, M. (1976). A rapid and sensitive method for the quantitation of microgram quantities of protein utilizing the principle of protein-dye binding. *Anal. Biochem.* 72, 248–254.

Eze, J. M. O. & E. B. Dumbroff (1982). A comparison of the Bradford and Lowry methods for the analyses of protein in chlorophyllous tissue. *Can. J. Bot.* 60, 1046–1049.

Lowry, O. H., N. J. Rosebrough, A. L. Farr & R. J. Randall (1951). Protein measurement with the Folin phenol reagent. *J. Biol. Chem.* 193, 265–275.

Kochert, G. (1978). Protein determination by dye binding. In J. A. Hellebust &

J. S. Craigie (eds.) *Handbook of Phycological Methods. Vol. 2. Physiological and biochemical methods.* Cambridge University Press, Cambridge, pp. 91–93.

Legette-Bailey, J. (1967). *Techniques in Protein Chemistry.* 2nd ed. Elsevier Publishing Co., New York.

Peterson, G. L. (1977). A simplification of the protein assay method of Lowry *et al.* which is more generally applicable. *Anal. Biochem.* 83, 346–356.

– (1979). Review of the Folin phenol protein quantitation method of Lowry, Rosebrough, Farr and Randall. *Anal. Biochem.* 100, 201–220.

– (1983). Determination of total protein. *Methods in Enzymology,* 91, 95–118.

Rausch, T. (1981). The estimation of micro-algal protein content and its meaning to the evaluation of algal biomass. I. Comparison of methods for extracting protein. *Hydrobiologia,* 78, 237–251.

NOTES FOR INSTRUCTORS
Plant material

Macrophytes collected in the early spring are easier to grind and contain more protein than material collected during summer. Therefore, it is recommended that this lab be conducted in the spring, if possible. This experiment can probably be conducted with other macrophytes, although some species have compounds that cause interference. For example, potassium, phenolics and some carbohydrates interfere with the Lowry method, while polysaccharides may interfere with the Bradford technique. Brown seaweeds such as *Fucus* have been found to show lower protein than one would expect from particulate nitrogen determinations.

Regarding the grinding procedure, if too much water is added, it is almost impossible to grind the tissue. On the other hand, if not enough water is added, the algal tissue will stick to the pestle. The mortar, pestle and glassware should be rinsed thoroughly with distilled water, as detergents such as Triton X-100, Haemosol, etc., will interfere with the color development. Grinding in boiling water is a very efficient way of extracting soluble protein, but it extracts other compounds, which may interfere with the Folin reaction.

Materials for each set

1. 2 g fresh wt of macrophyte tissue
2. mortar and pestle (porcelain or glass)
3. 1 g sand
4. 15 mL centrifuge tubes (2)
5. gloves and goggles
6. 10 mL test tubes (22) and test tube rack
7. pipettes (5 mL) (2)

8. 100 mL graduated cylinder (2)
9. Pasteur pipettes with bulbs (2)
10. volumetric flask (1 L) (1)
11. cuvettes (1 or 10 cm length) (2) (Volumes given are for small cuvettes)

Instruments

1. vortex mixer
2. spectrophotometer to read 750 nm
3. clinical-centrifuge (high speed $\sim$30,000 $\times$ g may be necessary if solution does not become clear after centrifuging)

Reagents

1. 2% Na_2CO_3 in 0.1 N NaOH (Reagent A)
2. 0.5% $CuSO_4 \cdot 5 H_2O$ in 1% sodium citrate (Reagent B)
3. 1 mL of reagent B in 50 mL of reagent A (Reagent C)
4. Folin-Ciocalteu reagent (dilute to 1 N) (Reagent D). Pour in fume hood. Wear goggles and gloves.
5. bovine serum albumin (50 mg 100 mL^{-1}) (e.g., Sigma Chemical Co.)
6. 100 mg Coomassie Brilliant Blue G-250
7. 50 mL 95% ethanol
8. 100 mL 85% H_3PO_4

Reagent preparation and storage

1. Coomassie Blue reagent. Dissolve 100 mg Coomassie Brilliant Blue G-250 in 50 mL of 95% ethanol. Add 100 mL 85% (w/v) H_3PO_4. Dilute resulting solution with water to 1 L. Store at room temperature.
2. Reagent B for the Lowry method should not be more than one month old.
3. The 100 g L^{-1} of bovine serum albumin stock solution must be refrigerated if left over night.

Alternatives

This lab may be shortened by testing only one of the protein methods (e.g., Lowry method).

Comments

Lowry method. For enhanced detectability, micro-adaptations have been described that can be used with as little as 0.2 μg of protein. Thus, the Lowry method is 50 to 100 times more sensitive than the standard biuret assay. If no spectrophotometer capable of providing readings at 750 nm is available, other wavelengths in the range of 500 to 750 nm (e.g., Hg 578 nm) can be used, but of course with reduced detectability.

Many modifications of the original procedure have been proposed to cope with interfering substances; details and references can be found in Peterson (1979). However, unless one is dealing with solutions that either are devoid of interfering substances or are so precisely defined that the same amount of the interfering substances(s) can be incorporated into the reagent blank and into the standard samples as is present in the unknown sample, it is highly advisable to separate the protein from interfering substances by precipitation with trichloroacetic acid. This also serves to concentrate proteins from very dilute solutions. Trichloroacetic acid alone does not precipitate proteins reliably and quantitatively at low levels (1–25 μg) of protein. This difficulty can be overcome by the combined use of trichloroacetic acid and deoxycholate or trichloroacetic acid and yeast soluble ribonucleic acid (0.125 mg mL^{-1}). The latter method can also be applied to precipitate proteins quantitatively in the presence of detergents such as dodecylsulphate, digitonin, or sulphobetaines (Bergmeyer & Grabl 1983).

Measurements of protein in kidney bean leaves revealed that chlorophyll can cause a significant increase in the absorbance readings because it absorbs at a similar wavelength (about 640 nm) to the protein complex formed in the Lowry method (Eze & Dumbroff 1982). Trichloroacetic acid precipitation of the protein eliminated significant interference by chlorophyll. Also, eluting leaf tissue with 80%–90% acetone prior to extraction of protein increased protein yield significantly. However, washing the protein precipitate with 80% acetone to remove residual chlorophyll resulted in large protein losses. Chlorophyll interferes less with the Bradford method because readings are made at 595 nm (below the absorption peak of chlorophyll).

Bradford method. A number of dyes have been used to stain proteins after electrophoretic separation in gels. These staining procedures are very sensitive, so attempts have been made to devise methods for protein quantitation with dyes such as Xylene Brilliant Cyanine G, Amido Black, bromosulphalein, or Coomassie Brilliant Blue R-250. Although these procedures possess high sensitivity, they require separation of the protein-dye complex from unbound dye by filtration or centrifugation. However, very simple methods have been reported that use Coomassie Brilliant Blue G-250 or bromophenol blue.

Several common laboratory substances, such as Tris, acetic acid, 2-mercapto-ethanol, sucrose, glycerol, and EDTA, have small but detectable effects when present in the Coomassie Blue-binding assay. To correct for this interference, the blank and standard samples should contain the appropriate amounts of the interfering substance, providing its concentration is known. Serious interferences are caused by phenol (>1 mmol L^{-1}), urea (>2 mol L^{-1}), guanidine hydrochloride (6 mol L^{-1}), ampholytes, alkaline buffers, and detergents such as Triton X-100 or sodium dodecylsulfate (>1 mg mL^{-1}). If the protein concentration is high enough, the sample solution can be diluted in order to lower the concentration of the interfering substance to an innocuous value. Alkaline solutions must be neutralized before the assay. To remove the interference by dodecylsulfate, precipitation of the detergent with potassium phosphate was described.

The Coomassie Blue-binding assay should be performed in glass or plastic cuvettes, as the protein-dye complex tends to bind to quartz cuvettes. Blue cuvettes can be cleaned either by rinsing with concentrated glassware detergent, followed by water and acetone, or by soaking in HCl (0.1 mol L^{-1}).

It appears that the Bradford method, while providing a rapid test for the presence of proteins (e.g., in chromatographic fractions), should be viewed with some caution. Reliable application is currently limited to measurement of proteins or protein mixtures that have been carefully standardized with other methods, or to situations where only very relative information is required (Peterson 1979).

Extraction. To measure soluble protein (cytosol, as opposed to insoluble or "membrane bound") in algae, extraction in distilled water would likely result in an underestimate, because many proteins would be less soluble in the acidic solution (owing to the release of organic acids from the tissue).

Phytoplankton protein is usually extracted with 3% trichloroacetic acid (TCA). TCA precipitates protein, but not free amino acids and short peptides. In combination with grinding with glass fiber filters, it is a very effective means of breaking up phytoplankton cells. The completeness of extraction appears to depend on the ease of cell-wall breakage (Rausch 1981).

4

Spectrophotometric and fluorometric chlorophyll analysis

Nanette T. Sterman

Dept. of Biological Sciences, University of California, Santa Barbara, CA 93106, USA

4A. Spectrophotometric analysis

FOR THE STUDENT

Introduction

Pigments are responsible for converting light energy to chemical energy in all photosynthetic organisms. Although photosynthetic algae and terrestrial plants each have a complement of several different pigments, they all have one common pigment–chlorophyll *a* (chl *a*). This chlorophyll is the only pigment found in the photosynthetic reaction center and is the actual pigment responsible for drawing electrons from water to initiate the light reaction of photosynthesis. Chl *a* absorbs light primarily in the red and blue wavelengths of the visible light range. To increase the ability of plants to use a wide range of light for photosynthesis, plants have evolved other pigments called antennae pigments or accessory pigments, which absorb light energy in other visible wavelengths such as yellow and orange. Antennae pigments pass the energy they absorb to the reaction center chl *a* for use in photosynthesis. Each group of algae has its own complement of accessory pigments, as you will see by the end of this experiment.

Plant cells are able to synthesize more or less of each type of pigment, not necessarily in constant proportions, but dependent on light and nutrient availability. In this way, plants maximize light harvest, that is, they capture the most light in as large a range as possible.

In this experiment you will take a known number of phytoplankton cells, break up the cells and the chloroplast thylakoid membranes (where the pigments are located) in a solvent. The color density of the resulting pigment extract will then be measured by a spectrophotometer. The spectrophotometer measures the amount of light that pigments absorb over the range of wavelengths of visible light. Each pigment has characteristic wavelengths at which absorbance will peak and from the location of the peaks, you will identify each pigment. From the height of the peak and the amount of absorbance, you will then calculate the amount of each pigment type extracted from the cells. Because you will know how many cells you extracted, you will be able to calculate the amount of each pigment per cell. These kinds of measurements are important for estimating biomass as well as determining the physiological condition of cells, such as adaptation to low or high light and general age of the population.

If you are making these measurements on a field sample, you will find that there are too many different kinds of plankton present for an accurate cell count. In this case, filter a known *volume* of seawater through a filter and extract the pigments from the filter. Your results will be on a volume basis rather than on a cell basis.

Procedure

1. For cultures, count cells (see Exp. 1) and calculate the number of cells per mL. Record this number.
2. Centrifuge 40 mL of a dilute culture in a 50 mL centrifuge tube, or 10 mL of a dense culture in a 15 mL centrifuge tube. Your instructor will help you determine the density of your culture. The following centrifuge speeds are for a tabletop clinical centrifuge and will yield best results:

Dinoflagellates	speed 5	2 min
Diatoms	speed 3	3–6 min
Green algae	speed 3	5–10 min

When the cells are pelleted and the supernatant fairly clear, proceed to Step 3. Note: If your cells are very small, they will not spin out easily. In this case, you will need to save a sample of the cells remaining in the supernatant after the centrifugation. Count these cells and subtract their number from the initial cell count to find the number of cells used in your extraction.

3. *Gently* pipette or aspirate off as much of the supernatant as you can without disturbing the pellet. With a little practice, you will be able to leave a nearly dry pellet. Discard the supernatant unless you need to count remaining cells, as described at the end of Step 2.

4. Using a small weighing spatula, add a pinch of $MgCO_3$ and 1 mL 90% or 100% acetone, depending on the type of organism (see following equations) to the pellet. Use a Pasteur pipette to resuspend the cell pellet and $MgCO_3$ in the acetone.
5. Dim the lights in the laboratory. Use the same Pasteur pipette to transfer the above mixture to an iced tissue homogenizer and grind the suspension using at least 15 strokes. Record the number of strokes you used and use the same number every time you grind the extract. *Caution!* Homogenizers are fragile; too much force will break them.
6. Use the Pasteur pipette to transfer the homogenate to a clean 15 mL centrifuge tube labeled "working extract." Replace tissue homogenizer in foil-wrapped 500 mL beaker of ice. Cover beaker with a foil lid.
7. Centrifuge the homogenate at Speed 5 for 30 sec, or until supernatant is clear. This supernatant is your pigment extract. Be careful not to spill a drop!
8. Use a clean pipette to transfer the pigment extract to a foil-wrapped 15 mL centrifuge tube that is marked with 0.1 mL calibrations and labeled "final extract." BE SURE TO KEEP ALL PIGMENT EXTRACTS ICED AND IN THE DARK; light bleaches the dissolved, extracted pigments.
9. The pellet of $MgCO_3$ contains broken cells, including fragmented chloroplast membranes that still contain pigments. Depending upon how dense and how old your culture is, Steps 5–8 will be repeated 4–10 times for a complete extraction. Following each homogenization and centrifugation, add the new clarified supernatant to the pigment extract, which should always be kept iced and dark. Keep the acetone volume as small as possible. Once the acetone supernatant is colorless, repeat 5–8 a final time.
10. When you are done extracting pigments, your test tube should contain approximately 10 mL of pigment extract. Unwrap the test tube and centrifuge the pigment extract to remove any residual $MgCO_3$. Record the volume of the extract, subtracting out the volume of $MgCO_3$. Rewrap and ice the pigment extract.
11. Your instructor will demonstrate use of the spectrophotometer for determining the optical density of the pigment extract. Use the wavelengths in the equations below for your algal type. Then, use the calculations to convert the absorbance values to chlorophyll concentrations.
12. If you are also using the fluorometer to determine the chlorophyll concentration, you will need to save at least 6 mL of pigment extract so that you can make serial dilutions. Keep all solutions dark and iced!
13. If you are working with a field population, you will need to prepare the cell sample differently. First, you will need to filter several liters

of seawater through a glass fiber filter. Filter enough water so that the filter looks pigmented—all the cells are being collected on the filter. Record the volume of the water filtered. Place the filter in a foil-wrapped 15 mL centrifuge tube and add a pinch of $MgCO_3$ and 10 mL of acetone. Be sure that the filter is fully submerged in the acetone. Cover with Parafilm, and swirl the test tube to mix the $MgCO_3$. Place the test tube in the freezer or a covered ice bucket for at least one hour. After the hour, seal the test tube with your thumb and invert the test tube to mix. Centrifuge the pigment extract and read it in the spectrophotometer, as per Steps 11 & 12. Use the appropriate calculation below.

Calculations

Calculate the amount of chl a cell^{-1} and chl b cell^{-1} or chl c cell^{-1} using the calculations that follow. First calculate μg chl mL^{-1} for each chlorophyll in your extract using the appropriate set of equations (1–4). Final equations give concentrations of chlorophylls in μmol cell^{-1} and μg cell^{-1}.

1. Higher plants and green algae containing chl a and b (solvent 90% acetone):

$$\text{chlorophyll } a = 11.93\ A_{664} - 1.93\ A_{647}$$
$$\text{chlorophyll } b = 20.36\ A_{647} - 5.50\ A_{664}$$

2. Diatoms, chrysomonads and brown algae containing chlorophylls, a, c_1, and c_2 in equal proportions (solvent 90% acetone):

$$\text{chlorophyll } a = 11.47\ A\quad - 0.4\ A$$
$$\text{chlorophyll } c_1 + c_2 = 24.36\ A_{630} - 3.73\ A_{664}$$

3. Dinoflagellates and cryptomonads containing chlorophylls a and c_2 (solvent 100% acetone):

$$\text{chlorophyll } a = 11.43\ A_{663} - 0.64\ A_{630}$$
$$\text{chlorophyll } c_2 = 27.09\ A_{630} - 3.63\ A_{663}$$

4. Mixed phytoplankton populations containing all of the above (solvent 90% acetone):

$$\text{chlorophyll } a = 11.85\ A_{664} - 1.54\ A_{647} - 0.08\ A_{630}$$
$$\text{chlorophyll } b = -5.43\ A_{664} + 21.03\ A_{647} - 2.66\ A_{630}$$
$$\text{chlorophyll } c_1 \text{ and } c_2 = -1.67\ A_{664} - 7.6\ A_{647} + 24.53\ A_{630}$$

Calculation for μg chl cell^{-1} of phytoplankton:

$$\mu\text{g chl in extract} = (\text{Volume of extract, mL})\ (\mu\text{g chl mL}^{-1})$$

$$\mu\text{mol of chl in extract} = \frac{\mu\text{g chl in extract}}{\text{molecular weight of chl}}$$

(Molecular weights: chl *a* 894, chl *b* 908, chl *c* 610.)

$$\mu\text{mol of chl cell}^{-1} = \frac{\mu\text{mol of chl in extract}}{\#\text{ cells in sample}}$$

If you do this experiment over the growth curve of a population, plot the following:

concentration of chl *a* cell^{-1}, chl *b* cell^{-1} or chl *c* cell^{-1} versus time

the ratio of chl *a* cell^{-1} to chl *b* cell^{-1}, or chl *c* cell^{-1} versus time

Questions

1. What kind of pattern do you see over time?
2. At what cell concentration did your culture reach the stationary phase?
3. How did the chlorophyll concentration per cell correspond to the growth curve?
4. Do the antennae pigments vary with or inversely to the chl *a* concentration?
5. When is chl *a* to antennae pigment ratio highest?
6. How do your results compare to those of other groups who used different types of algae?

REFERENCES

Hansmann, E. (1973). Pigment analysis. In, J. R. Stein (ed.) *Handbook of Phycological Methods. Culture Methods and Growth Measurements.* Cambridge University Press, Cambridge. pp. 359–368.

Jeffrey, S. W. & G. F. Humphrey (1975). New spectrophotometric equations for determining chlorophylls *a*, *b*, c_1 and c_2 in higher plants, algae and natural populations. *Biochem. Physiol. Pflanzen.* 167, 191–194.

Strickland, J. D. H. & T. R. Parsons (1972). *A Practical Handbook of Seawater Analysis* Fish Res. Bd. Canada Bull. 167, pp. 185–199.

NOTES FOR INSTRUCTORS

Materials for 12 pairs of students

Laboratory cultures.

For counting cells, see supplies listed for Experiment 1.

1. clinical centrifuges (tabletop) (3)
2. 40 mL glass centrifuge tubes (24)
3. 15 mL calibrated glass centrifuge tubes, marked in units of 0.1 mL (36)
4. pieces of aluminum foil large enough to wrap a 15 mL centrifuge tube and a 500 mL beaker (including lid) (12)

5. Pasteur pipettes (36, or more)
6. pipette bulbs (24)
7. glass tissue homogenizers (12)
8. 90% acetone (4L)
9. 100% acetone (1L)
10. 0.5 g $MgCO_3$
11. kimwipes
12. 500 mL squirt bottles filled with the 90% acetone (3)
13. 500 mL squirt bottles filled with the 10% acetone (3)
14. spectrophotometers each with 2 1-mL glass cuvettes (3 or more)
15. sink aspirators (2 or more)
16. ice in an ice chest
17. 500 mL plastic beakers for holding ice and test tubes (12)
18. marking pens (12)
19. 10-mL pipettes (12)
20. test tube racks (12)

Field sample. All of the equipment described for laboratory cultures, as well as:

1. vacuum pumps (2)
2. Millipore-type filter apparatus including tower, filter, clamp (4)
3. 4-L filtration flasks (4)
4. 1-L flasks (to be used as traps)
5. vacuum tubing and connectors to connect everything (2)
6. Y connectors for tubing (2)
7. filter forceps (4)
8. Gelman GF/C glass fiber filters or similar type of filter (12)

Connect pump, flasks, and tubing so that two 4-liter flasks are attached to one 1-L flask, (trap) which is attached to one vacuum pump.

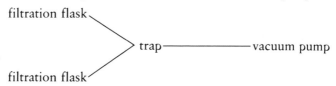

Comments

This procedure will take students 2–3 h the first time, 1 h after that. The procedure is most efficient if groups cooperate in using the centrifuge and are staggered for the spectrophotometers.

Small plankton are often difficult to centrifuge and filter completely, especially without bursting the cells. For example, *Chlorella* cells are extremely resistant to centrifugation. When your students work with these small plankton, have them count the cells in the initial sample, centrifuge the cells, then pipette off a small sample of supernatant before aspirating. Have them count the cells remaining in the supernatant and subtract that number from the initial cell count. The difference is the number of cells extracted.

Aspiration is an important step in the extraction. The idea is to remove as much medium as possible, because the water will change the acetone concentration and could change the wavelengths at which pigment absorbtions peak. Encourage students to end up with pellets that are nearly dry. It is actually quite easy to do without disturbing the pellet, but it requires some practice.

BE SURE STUDENTS KEEP EXTRACTS DARK AND ICED. Light fades pigment.

If your students repeat the chlorophyll extraction several times over the growth curve of a phytoplankton population, you will find that as the cell number increases, so does the time necessary for extraction. Therefore, have them cut back on the volume of the initial cell sample, i.e., from 40 mL down to 10 mL, as necessary. Be sure they always extract to completion.

Encourage students to use as little acetone as possible for each step of the extraction. Otherwise, they will end up with samples too dilute for the spectrophotometer to read, or they will have *two* centrifuge tubes of extract and no way to mix them.

The $MgCO_3$ acts as a buffer to delay chlorophyll breakdown.

For filtering the field sample, use gentle filtration, i.e., less than 380 mm Hg, otherwise the cells might burst.

All spectrophotometers operate somewhat differently. Show the students how to operate them. Remind students to hold the cuvettes on the frosted sides. Fingerprints on the clear sides interfere with the light path and can skew absorbance readings. Kimwipes and acetone are for wiping fingerprints off cuvettes.

4B. Fluorometric analysis

FOR THE STUDENT

Introduction

The fluorometer is another instrument for measuring chlorophyll concentration. This method has a number of advantages over the spectrophoto-

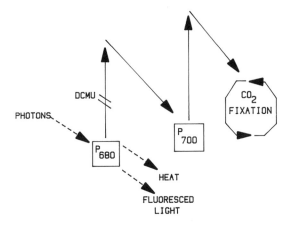

Fig. 4.1. Schematic showing (1) how light energy is converted to photosynthetic activity, fluorescence and heat; (2) where 3-(3,4-dichlorophenyl)-1,1-dimethylurea (DCMU) acts to disrupt photosynthetic electron flow between photosystem II (P_{680}) and photosystem I (P_{700}).

metric technique, such as having a higher sensitivity, requiring less concentrated samples, being fast, and direct. The fluorometer itself is small, portable and simple to use. The major disadvantage of fluorometry is that it measures only chl *a* and no accessory pigments.

Chlorophyll fluorometry operates on the following principle. The fluorometer emits short wavelength light that is absorbed by chlorophyll molecules and re-emitted (fluoresced) at a longer wavelength. The intensity of the fluorescence is measured by the fluorometer. From the fluorometer reading, you can calculate the pigment concentration.

Light absorbed by whole cell chlorophyll is used in three ways. Most is used for photosynthesis, some is emitted as fluorescence and the remainder is emitted as heat. Whole cells can be induced to emit all of the absorbed light energy as fluorescence by addition of DCMU (3-(3,4-dichlorophenyl)-1,1-dimethylurea). DCMU is a photosynthetic inhibitor that breaks the connection between photosystem II and photosystem I. All light energy absorbed by photosystem II is then fluoresced rather than being converted to chemical energy. With DCMU, you can determine a ratio of fluorescence by whole cells with intact photosynthetic pathways (F) to fluorescence by whole cells with photosynthetic pathways disrupted by DCMU (FDCMU). This ratio indicates the proportion of light absorbed by normal cells that is naturally fluoresced, another diagnostic for determining whole cell physiology.

When chlorophyll pigments are extracted from chloroplast thylakoid membranes, all absorbed light is fluoresced, as there no longer is an intact electron transport system. Extracted chl *a* absorbs light at 663 nm and fluoresces at 670 nm. Chl *a* in intact or whole plant cells absorbs light near 673 nm and 683 nm, and fluoresces that light at 685 nm.

Procedure for extracted chlorophyll

Serial dilutions. Since the fluorometer is more sensitive than the spectrophotometer, samples prepared for spectrophotometry must often be diluted for measurement by fluorometry. For this experiment you will first make serial dilutions of the chlorophyll prepared for spectrophotometric analysis.

1. Foil wrap and number five 10 mL test tubes with #1 being your starting solution and therefore the most concentrated sample.
2. Always keeping your extracted chlorophyll samples iced and dark, carefully measure at least 6 mL of extract into test tube #1 and record the volume.
3. Take 10% of the volume from test tube #1 and put it into test tube # 2. Add enough 90% or 100% acetone (depending on your algal type) to bring the volume up to the *initial* volume of test tube #1. Mix. You now have a 10% solution.
4. Take 10% of the volume from test tube #2 and put it into test tube # 3. Add acetone to test tube #3 to bring it to the initial volume of test tube #1. This makes test tube #3 a 1% solution.
5. Continue this process to make 0.1% and 0.01% solutions in test tubes #4 and #5, respectively.

Fluorometry

1. Your instructor will show you how to operate and calibrate the fluorometer. Then you must correct for the optical properties of the cuvette glass and the acetone. With a glass marker, mark a line 5 mm long from the rim of your cuvette. Each time you insert the cuvette into the fluorometer, be sure the cuvette is lined up in the same orientation by using the mark as your reference. Fill the cuvette with 90% or 100% acetone and measure the fluorescence. Record this number and the "door" you used. The different doors alter the characteristics of light produced by the fluorometer. The door factor is important in calculating the chlorophyll concentration of your sample. Acetone fluorescence is another correction factor for your calculations. It is *very* important that you measure a new acetone value each time you change the door and record both the acetone value and the door.
2. Discard the acetone and measure your pigment extracts. Pipette the most dilute pigment solution (0.01%) into the cuvette and measure its

fluorescence. Record the fluorescence value and the dilution. Fluorescence values above 50 are significant. If this concentration is too dilute (i.e., measures below 50), pour it back into its test tube and measure the sample that is ten times as concentrated (0.1%). Measure increasing concentrations until you get a significant fluorescence reading.

Procedure for whole cell fluorescence

1. Number four 10 mL test tubes.
2. Place 9 mL of well-mixed culture into each test tube. Your instructor will help you determine if your culture is dense enough for cells to need dilution. In dense cultures, cells shade each other and interfere with the fluorometer's ability to detect fluoresced light. If necessary, dilute the cells with filtered fresh water or sea water. Note the percent of your dilution.
3. Place the four test tubes in the dark for 15 min for the cells to dark-adapt.
4. While the cells are adapting to the dark, calibrate the fluorometer and the cuvette. Since the cells are in medium, measure the fluorescence of the cuvette filled with medium instead of acetone. Record the fluorescence.
5. After the 15-min dark adaptation, add 50 μL of DCMU to two of the test tubes. *CAUTION!!* DCMU is a poison so do not drink it and be sure to wash your hands! Record the numbers of the treated tubes, cover with parafilm, invert the tubes to mix, and replace them in the dark. The DCMU takes approximately 5 min to work. In the meantime, measure the fluorescence of the cells in the other two test tubes.
6. You will see the fluorometer dial move up and back for the first 45 sec after you insert the sample. This variation is due to fluorescence transients characteristic of whole cells. For the first sample you measure, record the time from when you closed the door until the fluorometer stabilizes, and record the fluorescence. Be consistent in the amount of time you wait for reading subsequent samples.
7. Use the calculations below for determining the chlorophyll concentration of the extracted solution, the whole cells, and the whole cells with DCMU.

Calculations

Final equations give concentrations of chl a in μmol chl a cell^{-1} and μg chl a cell^{-1}.

1. Multiply your chlorophyll fluorescence values by the reciprocal of the dilution factor you used.

2. μg chl a L^{-1} =

$$= \left(\begin{array}{c} \text{Fluorescence value} \\ \text{of pigment} \end{array} - \begin{array}{c} \text{fluorescence value} \\ \text{of blank} \end{array} \right) (\text{door factor})$$

3. μg chl a cell^{-1} = $\dfrac{(\mu\text{g chl } a)}{\text{L}} \dfrac{(1 \text{ L})}{1,000 \text{ mL}} \dfrac{(1 \text{ mL of initial sample})}{\#\text{cells mL}^{-1}}$

4. μmol chl a cell^{-1} = $\dfrac{\mu\text{g chl } a \text{ cell}^{-1}}{\text{molecular weight of chl } a, 894}$

Questions

1. Are the results as you expected?
2. What effect did DCMU have?
3. How do the results compare with those of other types of cells? The article referenced below will help you interpret your results.
4. How do they compare to the spectrophotometric results?

REFERENCES

Flovacek, R. & P. Hannan (1977). *In vivo* fluorescence determination of phytoplankton chlorophyll *a. Limnol. Oceanog.* 22, 919–925.

Strickland, J. D. H. & T. R. Parsons (1972). *A Practical Handbook of Seawater Analysis, Fish. Res. Bd. Canada Bull.* 167, pp. 201–203.

NOTES FOR INSTRUCTORS

Materials for 12 student pairs

Extracted chlorophyll. Chlorophyll extracts from procedure 4A.

1. 90% acetone (2L)
2. 100% acetone (1L)
3. 1 mL calibrated pipettes (15)
4. 5 mL calibrated pipettes (15)
5. 10 mL calibrated pipettes (12)
6. 10 mL test tubes (70)
7. test tube racks (12)
8. ice in an icebucket
9. 500 mL beakers to hold ice and test tubes (12)
10. foil squares to wrap 15 test tubes and one 500 mL beaker including lid (12)
11. squeeze bottles filled with 100 mL of 90% acetone (3)

12. squeeze bottle filled with 100 mL of 100% acetone (1)
13. glass marking pens (4)
14. Turner 111 fluorometers with Turner F-4T5-B bulbs (2 or 3)
15. 4-mL cuvettes (one per fluorometer) (2 or 3)
16. kimwipes

Whole cell chlorophyll.

1. 10 mL test tubes (48)
2. test tube racks (12)
3. 10 mL calibrated pipettes (12)
4. bulb-type safety pipette fillers (12)
5. 1 mL micropipetter and tips (2)
6. 1 L filtered seawater
7. fluorometers (2 or 3)
8. 4-mL cuvettes (one per fluorometer) (2 or 3)
9. kimwipes
10. glass marking pens (4)
11. a dark place such as a drawer, cupboard, dark cloth, etc.
12. 1 mL 1×10^{-6} M DCMU (10)

Comments

This procedure takes about one hour if chlorophyll extracts have already been prepared in Section 4A. Try to stagger student groups so that there is not a long line at the fluorometers.

To make the DCMU solution, first make a stock solution of 1×10^{-4} M DCMU in ethanol. Then dilute the stock solution to 1×10^{-6} M with distilled water.

The biggest problems with this experiment are (1) students not keeping the samples in the dark, and (2) mixing up samples.

All fluorometers calibrate slightly differently. When you show students how to calibrate the fluorometer, remind them to hold the cuvette by the rim, as fingerprints can skew readings, and to always place the cuvette in the same position. Kimwipes and acetone are good for removing fingerprints.

Do not expect spectrophotometer and fluorometer estimates of chl *a* to be the same.

5

Water quality assessment

Maureen A. Leupold

Ward's Natural Science Establishment, Inc. 5100 West
Henrietta Rd., P.O. Box 92912, Rochester, NY 14692-9012,
USA

FOR THE STUDENT

Introduction

One of the many effects that organic wastes have on the aquatic environment is to influence the type of algae that will grow. A survey made by Palmer (1969) revealed that more than 1,000 algal taxa have been reported as pollution-tolerant forms. The five genera of algae most tolerant of organic pollution are: *Euglena, Oscillatoria, Chlorella, Scenedesmus,* and *Chlamydomonas.* Various species of each of these prefer polluted waters, which are high in organic material.

Organic "enrichment," or pollution, is measured by the Biological Oxygen Demand (BOD) of the water. This test describes how much oxygen is required for the biological oxidation (biodegradation) of the organic substances in the water. A high BOD indicates that the rate of oxygen removal by bacteria and protozoa is high. Algal species tolerant to organic pollution are significant in the recovery of a stream or lake because they add oxygen to the water during photosynthesis and incorporate organic and inorganic nutrients into their cells, thus removing pollutants from the water.

Algal indicators of organic pollution can be used as a pollution index. An alga can be used in this manner if there are 50 or more individuals per milliliter of water sample. Table 5.1 was designed by Palmer (1969) and lists 20 algal genea and their corresponding pollution values. The individual pollution index factors of the algae in the sample are totaled. A score

Table 5.1. Algal genera pollution index (Palmer 1969)

Genus	Pollution index	Genus	Pollution index
Anacystis	1	Micractinium	1
Ankistrodesmus	2	Navicula	3
Chlamydomonas	4	Nitzschia	3
Chlorella	3	Oscillatoria	5
Closterium	1	Pandorina	1
Cyclotella	1	Phacus	2
Euglena	5	Phormidium	1
Gomphonema	1	Scenedesmus	4
Lepocinclis	1	Stigeoclonium	2
Melosira	1	Synedra	2

of 20 or more for a sample is evidence of high organic pollution. Mid-range scores (15–19) indicate that the organic pollution is moderate and low scores (below 15) indicate probable absence of organic pollution. However, a moderate or low Palmer score is not conclusive evidence of no pollution, since some other factor (such as low pH) might be restricting species diversity.

For example, a survey of a lake revealed the following data:

Number of organisms		Pollution index value
75	Chlorella	3
105	Scenedesmus	4
90	Euglena	5
205	Closterium	1
10	Chlamydomonas	(insufficient number)
140	Oscillatoria	5
62	Peridinium	(not an indicator)
50	Ankistrodesmus	2
		total 20

The total pollution index for the lake would be 20, indicating high organic pollution. Sources of high organic pollution are raw sewage, dairy wastes, food processing wastes and other oxygen-consuming wastes.

In this exercise, you will determine the degree of organic pollution by examining water samples for algal indicators outlined in Table 5.1. The

counting techniques you will use are employed in many areas of limnological study.

Procedures

In order to count and score the algal genera, the algae must first be identified. Make wet-mount slides from your water sample. It is best to compare live and preserved organisms. Use a key to the algae, such as Prescott's (1962), for identification, or compare your organisms to illustrations
provided by your instructor. Make a list of all algae identified in the
sample.

The objective is to count each of the algal species which have been identified by Palmer as organic pollution indicators in a sample of known volume. Refer to Table 5.1 for the list of indicators.

Two procedures follow. The first gives a simple method useful for laboratories that are not equipped with counting cells for each student. The
second is standard procedure using a Sedgwick-Rafter counting chamber.

Graph paper method

1. A 0.1 mL volume of sample can be evenly covered by a 22 × 22 mm
   coverslip. It would be tedious and time consuming to count the entire
   contents of the sample. Instead, strips of known area are counted. To
   measure the diameter of the microscope field:
   a) Take a 2 cm square of graph paper that has 1 mm blocks, place it
      on a slide, and cover with a 22 × 22 mm coverslip.
   b) While viewing under low (10×) power, align the blocks so that a
      square just touches the edge of the field (Figure 5.1). Count the num-
      ber of blocks across the diameter of the field. Record the results and
      the magnification.
   c) Repeat for all objectives except 97 × oil immersion.
   Note: For more accurate results, the field can be measured with a stage
   micrometer, if available.
2. Swirl a sample to mix the contents evenly. Fill a 0.1 mL pipette and
   deliver 0.1 mL of sample to a clean glass slide. Carefully cover with a
   22 × 22 mm coverslip, making sure no fluid leaks out from the sides.
   Handle the slide carefully and do not touch the top of the coverslip.
   Place the slide on the microscope stage.
3. Count and record the number of algae of each genus in a strip. A strip
   is a rectangle covering the full length of the coverslip by the width of
   the microscope field (Figure 5.2). It is best to count "algal units" (i.e.,

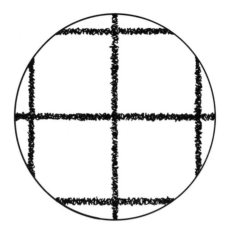

Fig. 5.1. Microscope field view of 1 mm graph paper at 100 power (10× objective × 10× ocular). Reprinted from Ward's Natural Science Establishment, Inc., Rochester, New York, Ward's Water Quality Assessment Kit, 1984.

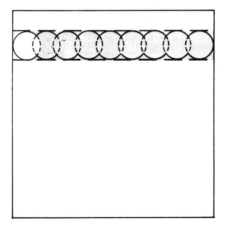

Fig. 5.2. A strip consists of the length of the coverslip (22 mm), with the width being the diameter of the microscope field. Reprinted from Ward's Natural Science Establishment, Inc., Rochester, New York, Ward's Water Quality Assessment Kit, 1984.

colonies and/or filaments are counted as one unit). Large filaments and colonies that are only partially lying in the strip should be counted as fractions (Figure 5.3). Repeat, counting at least three strips in total.

4. The density (number of algal cells or units per mL) is figured for each algal genus identified. Use the following formula to calculate cells or units per mL of concentrated sample:

Calculation A

$$\frac{\text{(Area of the Coverslip)} \times \text{(Number of units or cells for one algal genus)}}{\text{(Area of one strip)} \times \text{(number of strips counted)} \times \text{(the volume under coverslip)}}$$

For example:

The field of a microscope was measured using $100\times$ (10-power ocular with 10-power objective) and found to be 2 mm in diameter. Three strips of a slide holding 0.1 mL of sample under a 22×22 mm coverslip were counted. In total, 381.8 filaments of *Oscillatoria* were counted.

$$\frac{(22 \times 22 \text{ mm}) \times (381.8 \text{ } Oscillatoria)}{(2 \times 22 \text{ mm strip}) \times (3 \text{ strips}) \times (0.1 \text{ mL})} = 14{,}000 \text{ } Oscillatoria \text{ per mL sample}$$

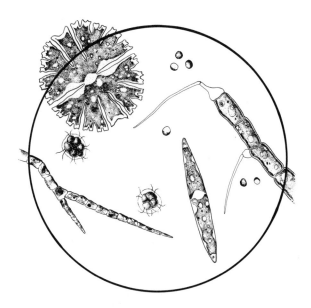

Fig. 5.3. Microscope field view of unpolluted water sample. Reprinted from Ward's Natural Science Establishment, Inc., Rochester, New York, Ward's Water Quality Assessment Kit, 1984.

This tells how many cells or units are contained in the *concentrated sample.*

5. The next step is to determine how many organisms are in the actual lake water. Use the following formula:

Calculation B

Concentration in lake water =

$$\frac{\text{(Number algae per mL concentrated sample)} \times \text{(total volume of concentrate)}}{\text{(Total volume of lake water filtered to yield concentrate)}}$$

In our example, 10 liters of lake water were filtered to yield a 100 mL concentrate.

$$\frac{\text{(14,000mL } Oscillatoria/\text{mL)} \times \text{(100 mL concentrate)}}{\text{10,000 mL lake water filtered}}$$

$$= 140 \text{ } Oscillatoria \text{ per mL}$$

Since *Oscillatoria* density is greater than 50 units/mL the sample would receive a score of 5. Each algal genus is counted, calculated and scored. The Palmer Score of organic pollution is totaled after all genera have been counted. Refer to the example in the introduction.

Note: Motile cells are difficult to count. A few drops of Lugol's iodine solution can be added to a small amount of the sample after observing the living cells.

Sedgwick-Rafter counting cell method. The Sedgwick-Rafter (S-R) counting cell is a device 50 mm long by 20 mm wide by 1 mm deep (American Public Health Association 1976) (refer back to Fig. 1.2). It has a 1,000 mm² area and holds a volume of 1 mL. The cell is filled, and strips of known area are counted.

1. Place the cover glass at a diagonal on top of the cell. Swirl the concentrated algal sample. Fill the cell with sample, using a large bore pipette, being careful to exclude any air bubbles. Do not overfill. Let the cell stand for about 15 minutes to allow the algae to settle.
2. A Whipple grid is a measuring device that is accurately ruled into 100 squares (American Public Health Association 1976). One of the squares is further divided into 25 smaller squares. The grid is placed in the eyepiece of a microscope. The measurements of the grid are calibrated to the microscope by measuring with a stage micrometer. Place a Whipple grid in the eyepiece of the microscope. Measure the width of the entire field of the Whipple grid by using a stage micrometer, while using the 200× objective. Record the width.
3. The algae will be counted in strips of known area and volume. A strip is the width of the Whipple grid field and the length of the S-R cell (50

mm), which has a depth of 1 mm. All algae in a strip are recorded as described earlier.

4. Use the following formula to calculate the number of algal units per mL.

Calculation C

Algal units per mL of concentrated sample =

$$\frac{(\text{number of algal units}) \times (1000 \text{ mm})}{(\text{strip length}) \times (\text{width of Whipple grid}) \times (\text{depth 1 mm}) (\text{number of strips counted})}$$

Determine the number of algal units per mL of lake water sample by using Calculation B.

5. Proceed with the scoring procedure outlined in the Graph Paper Method or Sedgwick-Rafter Counting Cell Method to determine the degree of organic pollution in your water samples.

Questions

1. The Palmer score indicates the degree of organic pollution by using algae as indicators of water quality. What other procedures are used to test for organic pollution? Based on your scores, what other test would you recommend to examine your water source?
2. If the Palmer score is low, does this indicate that the water is not polluted? Why or why not?
3. How are other organisms used as biological indicators?
4. What sources of error could occur with this counting procedure? Consider how your sample was obtained, the magnification used to count the algae samples, the preparation of your slides or counting chamber, and other sources of error.
5. If your Palmer scores were high, can you determine the source of organic pollution in the water body?

REFERENCES

American Public Health Association, American Water Works Association and Water Pollution Control Federation. *Standard Methods for the Examination of Water and Wastewater.* (1976). 14th ed. American Public Health Association, Washington DC. pp. 1007–1029.

Palmer, C. M. (1969). A composite rating of algae tolerating organic pollution. *J. Phycol.* 5, 78–82.

Prescott, G. W. (1962). *Algae of the Western Great Lakes Area.* 2nd ed. Wm. C. Brown Co., Dubuque, Iowa.

Wetzel, R. G. & G. E. Likens. (1979). *Limnological Analyses.* W. B. Saunders Co., Philadelphia, PA.

NOTES FOR INSTRUCTORS
Materials

1. compound microscope with mechanical stage (one per student)
2. concentrated lake water samples from different sources suspected of pollution
3. Lugol's solution in dropper bottles

   (60 g KI and 40 g I in 1 L distilled water) (one per student)
4. Identification keys to the algae and algal illustrations

5. microscope slides (one box)	OR	Sedgwick-Rafter (S-R) counting cell (one per student)
6. 22 × 22 mm coverslips (one box)		Whipple grid (one per student)
7. graph paper (one sheet)		Stage micrometer (one per student or group)
8. 0.1 mL pipettes (one per student)		Large bore pipette (one per student)

This exercise was developed for Ward's Natural Science Establishment (Ward's Natural Science Limited, Mississauga, Ontario, phone 416-279-4482; Ward's Natural Science Establishment, Rochester, New York, phone 716-359-2502). A kit by the same name is available from Ward's (catalog number 88 W 3056), which contains simulated water samples, an algal picture key, Student Work Sheet, and Teachers Guide.

Scheduling

The exercise can be conducted in one three-hour lab section if the Ward's Natural Science kit is used. The simple key to the algae in the kit speeds up the identification process.

If natural waters are the source of samples, they should be collected prior to this exercise. Students should also be familiar with the use of algal keys before proceeding with counting and scoring. A collecting trip and identification lab could be conducted in the laboratory time during the week preceding this exercise. The students could then examine living specimens and preserve them for the following week.

The class may be divided into groups and a sample assigned to each group. Total class counts can then be pooled to determine the final Palmer scores of the samples.

Comments

Algae collecting. A tow net with a metering device pulled through surface waters concentrates organisms while keeping track of the amount of water filtered. Algae can also be concentrated by pouring measured amounts of lake water through a stationary net. The concentrated samples are easier to analyze and yield statistically significant results. Where plankton densities are high, 1–2 L of water are sufficient for filtering; 6 L of water should be filtered where densities are low. The collected, concentrated samples should be adjusted to 100 mL.

The plankton samples can be preserved if examination of living cells is not possible. Lugol's solution is added at a concentration of 1 mL per 100 mL sample (American Public Health Association 1976). Lugol's will preserve most delicate species, but the organisms will be stained by the iodine. Store preserved samples in the dark.

6

Selenastrum capricornutum Printz algal assay: bottle test for determination of limiting nutrient status

David J. Rawlence

Division of Sciences, University of New Brunswick, Saint John, NB, Canada, E2L 4L5

FOR THE STUDENT

Introduction

Bioassays are broadly used in biology for tests as divergent as determining the influence of new drugs on humans, to gauging the effect of metal ions on aquatic ecosystems, or determining the element(s) limiting primary productivity in a particular lake or stream. Although these various tests all differ in detail from one another, there are certain elements in common:

1. All utilize an organism that is either relatively common, or that can be raised cheaply under laboratory conditions. Thus, animals such as white mice, rats, and rabbits are extensively used in drug testing, and algae are used routinely in a range of tests broadly concerned with water quality.
2. Bioassays generally utilize a group of organisms with relatively well-defined physiological characteristics. It is clearly necessary to interpret the results of any bioassay, and this is greatly aided by the existence of a comprehensive body of information about the physiology of the test organism.
3. Test results are recorded in terms of a small number of fundamental responses; growth and abnormal growth (tumor formation) are perhaps the most widely used parameters.

56

4. To facilitate the comparison of results between laboratories, most tests are carried out within a set of closely defined environmental parameters. The normal growth of the test organism within these conditions is generally known. In this way, with the application of one of a number of simple statistical tests, it is relatively easy to determine which results differ significantly from the control, and to compare results between one laboratory and another.

Within the biosphere, freshwater constitutes a finite resource. The continuing expansion of the human population is a growing stress on this fixed resource. Almost daily, new questions are raised about the effect of a particular stress (for example an industrial discharge) on a particular freshwater system, or about the effect of a new stress on freshwater supplies in general. For example, the effect of continuing acidification on lakes in many regions is of general concern to governments worldwide. Yan & Stokes (1978) conducted *in situ* experiments, using natural phytoplankton assemblages, to determine the effect of pH manipulation on the phytoplankton of lakes in midwestern Canada. More specific questions may also be addressed. For example, Weber (1981) used *Chlorella* to screen water samples for the presence of metals, pesticides and other toxic substances.

Answers to these kinds of questions are fundamental to the wise management of lake and river catchments. The *Selenastrum capricornutum* test given here is a version of that routinely used in many research laboratories to provide both general and specific answers to a broad range of questions concerned with the management of freshwater resources.

Primary production is rarely, if ever, free of all physical and chemical constraints. Liebig first expressed the idea that there were always chemical constraints to growth, in the form of his "Law of the Minimum." Stated simply, the essential element present in the smallest quantity is that which governs growth.

The determination of which element limits algal growth is of special interest when deliberate changes in the quality of a particular body of water are contemplated. The potential influence of treated or untreated sewage wastes or industrial discharges on a system is normally assessed by determining which element naturally limits primary productivity. The bioassay described in this laboratory constitutes a widely used approach to determining the influence on primary productivity of a proposed change in water chemistry.

In its simplest form, the test is used to determine whether growth in a receiving water is naturally limited by nitrogen or phosphorus. The advent of an N-rich discharge to a body of water in which growth is naturally limited by N will clearly result in increased primary production. Such a development would probably be considered undesirable in an oligotrophic system utilized for recreational purposes, or in a potable water supply.

A variation in the test protocol may be used to determine whether growth in a system is naturally limited by the presence of heavy metals. In this way, for example, the influence of an existing or proposed industrial discharge on a system may be evaluated.

The algal assay described as follows is based on the widely used *Selenastrum capricornutum* Printz algal assay: bottle test developed by Miller et al. (1978). *Selenastrum capricornutum* (Chlorococcales) is a widely distributed alga, not commonly found in large concentrations. It is easily maintained under laboratory conditions, in which it retains a constant shape and discrete form (nonclumping). In addition, a great deal is known about its physiology (see Miller et al. 1978 for review).

The main elements in the test routine are:

1. Sterilization of the test water samples;
2. Inoculation of the test water samples with a standard number of *Selenastrum* cells;
3. The addition of sufficient nutrient(s) to support growth in the flasks to the level dictated by supply of the secondary limiting element;
4. Measurement of yield; and
5. Analysis of the data.

Procedure

1. Add sterilized water to 16 sterile gauze-stoppered flasks. Your instructor will tell you what volume to add to each flask.
2. Add 1 mL of the *Selenastrum* inoculum to each flask.
3. Make the following additions to each pair of flasks. (The volumes of each will vary according to the concentration of each of the stock solutions prepared and the volume of your test solutions. Your instructor will provide this information).

   Control (sterilized water sample)
   Control + 0.05 mg P L^{-1} as K$_2$HPO$_4$
   Control + 1.00 mg N L^{-1} as NaNO$_3$
   Control + 0.05 mg P L^{-1} + 1.00 mg N L^{-1}
   Control + 1.00 mg Na$_2$EDTA L^{-1}
   Control + 0.05 mg P L^{-1} + 1.00 mg Na$_2$EDTA L^{-1}
   Control + 1.00 mg N L^{-1} + 1.00 mg Na$_2$ EDTA L^{-1}
   Control + 0.05 mg P L^{-1} + 1.00 mg N L^{-1} + 1.00 mg Na$_2$ EDTA L^{-1}

4. LABEL ALL TREATMENTS CLEARLY.
5. Incubate your samples for the period recommended by your instructor. Take care to keep the test solutions off the gauze stoppers if agitating by hand.
6. Measure the fluorescence of each sample, as in Experiment 4B. If this experiment is being used to illustrate aspects of population growth, measurements should be made every 2–3 days.

Interpretation of results

As a guide, differences in yield of less than 20% are generally not significant. In P-limited waters, there should essentially be no difference between the yield of control + N, and N + EDTA treatments.

Results indicative of N-limitation are essentially the reciprocal of the above. Nitrogen and P co-limitation is not common. It is normally encountered in highly eutrophic or ultraoligotrophic conditions. Co-limitation is characterized by essentially similar yields in Control, +N, +P, and +EDTA (i.e., yields within 20% of one another).

Questions

1. Would you expect N or P limitation to be more common in freshwaters? Give reasons for your answer.
2. Is there a fundamental difference between the common limiting nutrient in freshwaters and that limiting primary production in the sea? If so, what is the basis for the difference?

REFERENCES

Miller, W. E., J. C. Greene & T. Shiroyama (1978). The *Selenastrum capricornutum* Printz algal assay: bottle test. Experimental design, application, and data interpretation protocol. EPA-60019-78-018. 124p.

Rehnberg, B. G., D. A. Schultz & R. L. Raschke (1982). Limitations of electronic particle counting in reference to algal assays. *J. Wat. Pollut. Control. Fed.* 54, 181–186.

Sellner, K. G., L. Lyons, E. S. Perry & D. B. Heimark (1982). Assessing physiological stress in *Thalassiosira fluviatilis* (Bacillariophyta) and *Dunaliella tertiolecta* (Chlorophyta) with DCMU-enhanced fluorescence. *J. Phycol.* 18, 142–148.

Weber, A. (1981). An uncomplicated screening test to evaluate toxicity of environmentally hazardous compounds in water. *Environ. Technol. Lett.* 2, 323.

Yan, N. D. & P. Stokes (1978). Phytoplankton of an acidic lake, and its responses to experimental alterations of pH. *Environ. Conserv.* 5(2), 93–100.

NOTES FOR INSTRUCTORS
Materials and equipment

1. Turner 111 Fluorometer equipped with a high sensitivity door.
2. R 136 red-sensitive photomultiplier
3. Corning blue CS 5-60 primary filter and red Corning CS 2-64 secondary filter.

Illumination: "Cool white" fluorescent light of 100 μE m^{-2}s^{-1} $\pm$ 10%

Temperature: Normally, 24°C($\pm$ 2 C), but this is not critical in a teaching situation. In a research setting, any irregular variation in growth rates due to temperature fluctuations would make comparison between different experiments difficult.

Automatic pipette: The use of an automatic pipette is recommended for the allocation of the *Selenastrum* inoculum. If none is available, the sterilized water sample may be inoculated with *Selenastrum*, and an appropriate volume added to each of the sterile flasks.

Filters: 0.08 μm Millipore or 1.2 μm Millipore (or glass fiber).

Source of *Selenastrum*: American Type Culture Collection, Sales Department, 12301 Parklawn Drive, Rockville, MD 20852.

Chemicals and glassware

Assorted pipettes.
16 125 mL and 500 mL Erlenmeyer flasks.
P-free detergent (for washing all glassware associated with P-limitation experiments).

Solution A.
  Dissolve in 500 mL of distilled water:
    12.750 g NaNO$_3$
    6.082 g MgCl$_2 \cdot$6H$_2$O
    2.205 g CaCl$_2 \cdot$2H$_2$O
    92.76 mg H$_3$BO$_3$
    207.69 mg MnCl$_2 \cdot$4H$_2$O
    1.64 mg ZnCl$_2$
    79.88 mg FeCl$_3 \cdot$6H$_2$O
    0.71 mg CoCl$_2 \cdot$6H$_2$O
    3.63 mg Na$_2$MoO$_4 \cdot$2H$_2$O
    0.006 mg CuCl$_2 \cdot$2H$_2$O
  150.00 mg Na$_2$EDTA

Solution B.
  Dissolve in 500 mL of distilled water:
    7.350 g MgSO$_4 \cdot$7H$_2$O

Solution C.
  Dissolve in 500 mL of distilled water:
    0.522 g K$_2$HPO$_4$

Solution D.
  Dissolve in 500 mL of distilled water:
    7.500 g NaHCO$_3$

Maintenance of the Selenastrum *stock*

The stock culture should be maintained in the log phase by subculturing approximately every 10 days. When pre- or post-log phase cultures are used for inoculation, a noticeable lag phase often results. Add 1 mL of solutions A through D to 900 mL of distilled water and make the volume up to 1 L with distilled water. Adjust the pH of the final medium to 7.5 $\pm$ 0.1 with 0.1 N HCl or NaOH. If an electronic particle counter is to be used for the determination of cell numbers, the final solution should be filtered through a 0.45 μm Millipore filter. Autoclave the solution and store in darkness at 4° C to avoid any photochemical changes if not prepared for immediate use.

Use of natural waters

Lake water samples should be sterilized by autoclaving, which releases some nutrients incorporated within algal cells. Water chemistry is altered as a result, the pH in particular being changed, and CO_2 released. After autoclaving and cooling, the sample is bubbled with 1% CO_2 in sterile air for 2 min L^{-1}. It is not a wise practice to store samples for more than a week prior to inoculation.

Water with total hardness in excess of 150 mg L^{-1} may produce an almost insoluble Ca and P precipitate, although the concomitant pH change may be relatively minor. Clearly the formation of such a precipitate may constitute an important source of error in some cases. For teaching purposes, the use of hard waters should therefore be avoided.

Preparation of inoculum

Selenastrum cells from the stock culture should be freed of medium by centrifugation. Between 15 mL and 20 mL should be centrifuged at 100 $\times$ g for 5 min. This should be repeated at least twice, resuspending the cells each time with distilled water. The concentration of cells in this sample should then be determined with the use of a counting chamber and microscope (with phase contrast illumination if possible) (see Experiment 2). The amount of the cleaned stock culture to add to the stock inoculum is determined with the following formula:

$$Q = \frac{A.\ B.\ C}{D}$$

where A = required inoculum volume (see following paragraph),

      B = desired *Selenastrum* concentration in flasks,

      C = volume of solution in test flasks,

      D = *Selenastrum* concentration in the cleaned stock,

      Q = volume of cleaned stock to add to each flask.

It is important to add the same number of cells to each flask. A larger volume of stock inoculum is normally prepared, as this allows a vigorous mixing of the stock during inoculation, which is very important if a large number of replicates is used.

As a guide, each test flask might conveniently contain 60 mL of test solution, if twice daily agitation of cultures is followed, rather than continuous agitation. This allows a reasonable surface/volume ratio in 125 mL flasks. For a class of five groups, with a total of 80 flasks, perhaps as much as 200 mL total inoculum should be prepared. The product of 200 × 1000 (cells mL) × 60 is divided by the cell count of the cleaned suspension to give the volume (Q). The volume (Q) is added to a volumetric flask and made up to 200 mL. One mL of this substock is added to each flask to give an initial cell concentration (B) of 1000 cells mL^{-1}.

Results

The results of these experiments may be expressed on the basis of dry weight, cell number determined by electronic particle counter, cell number derived from a fluorescence calibration curve, or cell number determined by counting chamber.

The advantages and disadvantages of the different techniques are discussed in Miller et al. (1978). Although Miller suggests the use of Millipore BD filters with 0.6 μm porosity for dry weight determination, I have not found the technique satisfactory.

Some workers (e.g., Miller et al. 1978) hold that results must be converted to dry weight, expressed as mg L^{-1}. This view is not universally held. Rehnberg et al. (1982) showed that in comparison with cell counts, phaeophytin *a* and chlorophyll *a* fluorescence measurements *in vivo* gave an accurate and rapid indication of toxicity to *Selenastrum capricornutum*. Algal fluorescence enhanced by DCMU (Exp. 4B) may form the basis of an almost instantaneous test eventually replacing the bottle test procedure described here (Sellner et al. 1982). To interpret the results of the algal assay:bottle test, it is not essential, therefore, to convert fluorescence readings to cell number.

If desired, the fluorometer may be calibrated by measuring the fluorescence of a serial dilution of a known concentration of *Selenastrum*. This may be simply achieved by centrifuging a few mL of healthy stock *Selenastrum,* determining cell numbers with a counting chamber, and making an appropriate dilution series from that sample of known concentration.

Conclusion

On occasion, the sterilization of water samples used for bioassays may result in significantly decreased growth, in comparison with that in Millipore-filtered controls. The reasons for this difference are not entirely clear. The choice between autoclaving and Millipore filtration is clearly a choice between relative freedom from contaminants versus somewhat enhanced growth. In the final analysis, some simple experimental work prior to this laboratory may suggest the most profitable experimental approach with the test waters available.

7

Photosynthesis and respiration of aquatic macro-flora using the light and dark bottle oxygen method and dissolved oxygen analyzer

Martin L. H. Thomas

University of New Brunswick in Saint John, Saint John, New Brunswick, E2L 4L5, Canada.

FOR THE STUDENT

Introduction

With few exceptions, energy used in the metabolism of all living organisms comes from the fixation of sunlight in the process of photosynthesis. This process is carried out by chlorophyll-containing plants, which convert inorganic carbon in the form of carbon dioxide or carbonates from their environment into organic compounds. Although the biochemical processes involved are complex, the process can be followed by reference to the basic equation for photosynthesis written thus:

$$6CO_2 + 6H_2O + \text{Light Energy} = C_6H_{12}O_6 + 6O_2$$

The process of respiration, which releases energy, is essentially the reverse of this reaction. It is possible to use any of the compounds, with the exception of water, to follow processes of photosynthesis and respiration.

Aquatic plants are an especially easy group to work with in following changes in oxygen in photosynthesis and respiration. In aquatic environments, the oxygen involved is normally in dissolved form readily measured with a dissolved oxygen analyzer. Finenko (1978) gives a thorough description of all aspects of aquatic productivity. An alternative titration technique, the Winkler method, is described in detail in the Appendix to Experiment 7.

Dissolved oxygen analyzers are available with electrodes of two different types, termed active and passive, and in many different models. All operate on the same principle: the probe contains a gold anode and a silver cathode in a solution of buffered potassium chloride. The silver cathode may be either visible or internal, but the anode is clearly visible on the probe tip. The central anode and surrounding cathode, or wicks to the cathode, are covered with a thin teflon membrane which separates them from the water. Oxygen from the water diffuses through the membrane, setting up a current between the anode and the cathode that is directly proportional to the percent saturation of the sample. This utilization of oxygen means that the sample must be continually replaced at the probe tip. This is accomplished by a stirrer (built in on many models, but an accessory on others). Without adequate stirring, the meter will show a steadily declining oxygen level. Because the response of the probe is to saturation of the sample rather than absolute oxygen concentrations, and saturation is temperature-dependent, analyzers carry out a correction based on sample temperature.

The light and dark oxygen method involves the use of two bottles—one clear and one dark—of known initial oxygen content and containing matched samples of parts, or small entire macro-plants, or just water containing plankton, which are incubated for a suitable time in the natural, or simulated, environment of the plants used. The plants or plant parts used will not be just a sample of the species chosen, but will include associated epiphytes; the water introduced into the bottles may contain natural phytoplankton, zooplankton, and bacteria communities. The results, then, will be representative of a community; however, for macrophytes, which represent the bulk of the response, the result will reflect the species chosen. There are some problems associated with the method (as discussed in greater detail), but generally it does yield ecologically useful results (see also reviews such as Vollenweider 1969; Thomas 1976; Littler 1979).

The situation in the dark bottle is relatively simple since all organisms should be respiring and no photosynthesis will occur. The loss of oxygen during incubation in the dark bottle should then represent respiration. In the light bottle, respiration will still be taking place, both from the contained micro-fauna and from the plants. In the case of the plants, however, under normal conditions of illumination, it would be expected that the evolution of oxygen in photosynthesis would exceed that used in respiration. The light bottle, therefore, should normally show an increase in dissolved oxygen, which is a measure of net production. Assuming that all cells in the bottle are mostly the selected plant or plankton, then the oxygen loss in the dark bottle added to the oxygen gain in the light bottle should give a measure of the total photosynthesis, or gross production.

Although the method is simple, drawbacks stem from the confinement of the sample in a relatively small bottle. The large surface area of the inside of the bottle provides a suitable substratum for the settlement and proliferation of bacteria, resulting in enhanced respiration. This problem can be controlled by using short incubation times, normally two to four hours, and by scrupulous attention to the cleanliness of bottles. Second, lack of free circulation of the water may allow a temperature increase in either bottle and may promote stratification, since mixing will be reduced or eliminated. A third set of problems involves the preparation of plant samples to be incubated in the bottles (Vollenweider 1969; Littler 1979). Most aquatic macrophytes are too large to incubate whole and therefore must be divided for incubation. Wounding and consequent loss of internal fluids will affect photosynthesis, and different plant parts will show differences in production and respiration. Nevertheless, the light and dark bottle oxygen method does yield reproducible, representative results and additionally it has the big advantage over the ^{14}C method of giving all three components: gross and net production and respiration.

The outcome of a standard experiment under typical conditions shows an oxygen decrease in the dark bottle and an oxygen increase in the light bottle. In this case a positive result will be obtained for gross and for net photosynthesis and respiration. A second frequent scenario sees a reduction in oxygen in the light bottle, but an even larger reduction in the dark bottle. Here, calculations will yield positive results for gross photosynthesis and respiration and a negative result for net photosynthesis. In this case, net photosynthesis should be given as zero, since, by definition, net production cannot have a negative value. In this second case, the primary producers are not producing enough oxygen to compensate for respiration in the bottle. Some production, however, is shown by the positive value for gross photosynthesis. An oxygen loss in the light bottle equal to that in the dark would indicate a total lack of production. In several thousand experiments of this type performed, I have seen this situation only once. It would suggest either that the plants in the bottle have just died, or that they are photosynthetically inactive. A greater loss in oxygen in the light bottle as compared to the dark bottle is impossible unless the weight of samples in the two bottles are very different or some unnoticed animal was included in the light bottle. At any rate, it would rarely allow the calculation of useful results. A gain in oxygen in the dark bottle is impossible under normal conditions. It could arise from the inclusion of a bubble of oxygen with the sample, quite possible in the case of tubular algae such as *Enteromorpha* spp. However, it usually means an experimental error has taken place. Whatever the cause, it means that dark bottle results cannot be used in calculations. In this case, only net photosynthesis values can be salvaged.

Although the individual experiments are simple and their immediate interpretation is straightforward, they are representative only of the incubation period. It is always tempting to extrapolate such experiments to a wider frame of time and space, such as per day. This can be done but the results obtained can only be general estimates unless a great deal of background work is done.

Procedure

Specimen selection. At the start of the laboratory, select test plant species and assemble specimens. For macro-algae, either select enough small whole plants so that very similar pieces can be introduced as pairs into light and dark bottles, or cut desired parts of the plant body. In many algae, the selection of almost identical samples is easy because of the habit of dichotomous branching. As a rough guide, specimens should be about 10cm long and/or weigh about 1–3 g. Overloading B.O.D. bottles with material will inevitably result in 'bubble trouble.' Selected samples should be placed in a pail of the water being used in the experiment until the bottles are prepared. The water to be used in the incubation bottles should be filtered through a 63 μm nylon screen and then bubbled with nitrogen if its natural oxygen content is above 50% of saturation.

Bottle filling. The filling of the experimental bottles and an 'initial' bottle (for O_2 starting concentration) is one of the most critical phases of the experiment. It is essential that all bottles be filled from a homogeneous water sample. Two methods achieve this: 1) In the simplest, B.O.D. bottles are filled by lowering them slowly into a container of water while tilted at an angle of 45°. Water must run in smoothly without any bubbling. As the bottle nears full, it is tilted slowly to the vertical to avoid trapping bubbles in the shoulder; and 2) Bottles may also be filled by use of a tygon tube installed as a syphon or from a spigot: the free end of the tube must start at the bottom of the B.O.D. bottle, remain there until the bottle overflow, and then be withdrawn while the water is still running slowly. This ensures that the last water sampled stays in the bottle and that the water that 'flushed' the bottle is discarded. Bottles should be filled to the top of the built-in funnel, tapped gently with the stopper to dislodge small bubbles and set to one side without stoppering. Fill two bottles for each experiment, plus one for initial oxygen.

Starting the experiment.

1. Place the samples in pairs of bottles and replace the stoppers, taking care not to trap any bubbles.
2. Record bottle numbers.
3. Cover dark bottle(s) with foil. Do not leave any gaps, even around the stopper.
4. Place bottles in a naturally or artificially lighted incubation vessel, in water, on their sides, with the number down in light bottles. Be especially careful to incubate light bottles in illumination typical for the species being tested.
5. Record time.
6. Read the oxygen concentration in the initial bottle and record it. Discard the contents of the initial bottle.

Incubation. Allow incubation to proceed 1–4 hours according to conditions. If bubbles start to appear in light bottles, terminate incubation.

Termination of the experiment.

1. At the finish of incubation remove the bottles from the incubation vessel, immerse the electrode, read and record the dissolved oxygen content on the analyzer. If bottles need to be moved a short distance to the instrument, do so in a bucket in which the bottles are immersed in water. Do not remove foil from dark bottles until they are ready for electrode insertion. If samples foul the stirrer, they must be freed. If bottles have bubbles, try to amalgamate them to one bubble, before opening the bottle, by holding the bottle at 45° and tapping it with a spare stopper. At least reduce the bubbles to a few. Measure the diameter of each bubble in mm while holding the bottle at 45°. The bubble(s) must be in the curved shoulder of the bottle for measurement and should be viewed at right angles (oblique observation would introduce parallax errors into the measurement).
2. Record bubble diameter(s) and dissolved oxygen concentrations for each bottle number.
3. Carefully pour the contents of the bottle away so that the specimen is retained.
4. Weigh the specimens from each bottle with surface water blotted off. Continue with the determination of dry weights if desired.

Calculations

The following abbreviations are used in the equations that follow. Oxygen levels are in mg L^{-1} (ppm).

IB = Initial Bottle oxygen level.
LB = Light Bottle oxygen level.
DB = Dark Bottle oxygen level.
N = Incubation time in hours.
PQ = The photosynthetic quotient. (Use 1.2)
RQ = The respiratory quotient. (Use 1.0)

Preliminary calculations. The following calculations (for $mgC\ m^{-3}\ h^{-1}$) are standard for phytoplankton experiments and are taken from Strickland and Parsons (1972). They are a necessary preliminary step for macroplant experiments but the results cannot be used directly.

Gross photosynthesis = 375.9 (LB − DB)/(N × PQ)
Net photosynthesis = 375.9 (LB − IB)/(N × PQ)
Respiration   = 375.9 (IB − DB) × RQ/N

The factor (375.9) converts the results to a standard volume (1 m^3).

Macro-plant experiments. If bubbles are present in the bottles, the following bubble correction is a required preliminary step before proceeding with calculations. For each bottle: 1) Determine the volume of gas in the bubble or bubbles, from their diameters, by reference to Fig. 7.1. 2) Subtract the total dark-bottle bubble volume from the total light-bottle bubble volume. 3) Convert the result (volume of oxygen in B.O.D. bottle) to mg $O_2\ L^{-1}$ by multiplying by 2.31. 4) Add the result to the dissolved oxygen analyzer reading for the *light bottle.* Then carry out the three preliminary calculations given above.

Call the three results: PG, PN and R respectively, then proceed with the following three calculations to yield results as $mgC\ g^{-1}$ organism h^{-1}.

Gross photosynthesis = PG/3333 (wt. sample g)
Net photosynthesis = PN/3333 (wt. sample g)
Respiration   = R/3333 (wt. sample g)

The factor (3333) standardizes results to a liter basis.

The weight of sample in the light bottle is used for photosynthesis calculations and the weight of sample in the dark bottle for respiration. Ideally, samples should be of identical weight.

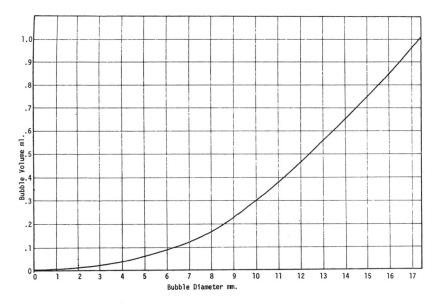

Fig. 7.1. Graph for the conversion of bubble diameter to volume in necks of 300 mL BOD bottles.

Alternatively, if a series of treatments was used, results may be compared on the basis of mg O_2 g^{-1} h^{-1}, instead of converting to carbon.

Questions

1. What is the energy-fixing process being investigated in this experiment? Give the basic equation describing the process.
2. Differentiate between net photosynthesis, gross photosynthesis and respiration.
3. What two processes are taking place in the light bottle and how does the dark bottle allow us to separate these?
4. What does a dissolved oxygen analyzer actually measure?
5. Why does a dissolved oxygen analyzer respond more slowly to temperature than to oxygen?
6. Why must a dissolved oxygen analyzer always be used with a sample stirrer?
7. Why are bubbles 'trouble'?
8. Would it be possible, under normal conditions, for an increase in oxygen to occur in a dark bottle?
9. If production is estimated for a whole day, why is the dark reading multiplied by 24, but the light reading by 8–16?

10. Would wet or dry weights provide the best basis for comparing photosynthesis between different species of plants?
11. How would you interpret a drop in oxygen level in a light bottle?

REFERENCES

Finenko, Z. Z. (1978). Production in plant populations. In, O. Kinne (Ed.) *Marine Ecology*, Vol. 4, *Dynamics*. John Wiley and Sons, New York, pp. 13–88.

Littler, M. M. (1979). The effects of bottle volume, thallus weight, oxygen saturation levels, and water movement on apparent photosynthetic rates in marine algae. *Aquat. Bot. 7*, 21–34.

Strickland, J. D. H. & T. R. Parsons. (1972). *A Practical Handbook of Seawater Analysis*. Bulletin #167, Fish. Res. Bd. Canada, Ottawa, Ontario, Canada 310p.

Thomas, M. L. H. (1976). *Introducing the Sea*. Huntsman Marine Lab, St. Andrews, New Brunswick, Canada. 112p.

Vollenweider, R. A. (1969). *A Manual on Methods for Measuring Primary Production in Aquatic Environments*. I.B.P. Handbook #12, Blackwell Scientific Publ. Oxford, U.K. 212p.

NOTES FOR INSTRUCTORS

Supplies

Bottles. For each basic experiment, three B.O.D. bottles are required. One for the initial oxygen sample and two for incubation. For series of experiments started together, you need one bottle for initial oxygen and two for each test.

Other supplies

1. Sufficient foil to cover all dark bottles.
2. Calibrated oxygen analyzer with sensor in a bottle of water (see as follows).
3. Two pails, one for water to stabilize probe and one for specimen preparation.
4. Scissors or scalpel for preparation of samples.
5. Watch or timer.
6. Notebook and pencils.
7. Salinometer. (A salinity refractometer or salinity conductivity meter is recommended in tidal pools. Hydrometers and induction salinometers are not satisfactory.)
8. An approximate 5 mm diameter piece of plastic tubing for a syphon if needed (see as follows).
9. A balance capable of weighing to 0.01 g (in laboratory).

The oxygen analyzer

The only special instrument needed is the dissolved oxygen analyzer. A passive electrode type is recommended over an active type. Suppliers follow.

Because the membrane must be in intimate contact with the sample, the thermistor is situated behind it. This distance from the sample means that it inevitably takes several seconds for the thermistor to fully respond to temperature. Failure to understand this process leads to the majority of errors in oxygen measurement. Readings will inevitably drift until the probe reaches sample temperature. The time delay will be proportional to the temperature difference between the probe and sample. A few simple precautions can lessen problems from this source. The probe should be stored, prior to readings, in a bottle kept at about sample temperature. In practice, a pail of water with a bottle for the probe immersed in it works well. If many samples have to be read in a short period, speed and accuracy can be maximized by grouping samples of similar probable oxygen and temperature levels. In any event, a drifting reading will not give accurate results.

Two types of analyzer, the active and passive, were mentioned earlier. In the passive type, such as instruments made by Delta-Xertex or Orbisphere, a constant voltage must be applied to the electrode whether or not it is in use. In passive instruments, the electrode must not be disconnected from the read–out unit during periods of use. These instruments energize the probe even when they are turned off. If the electrode is disconnected, at least 30 min must elapse between reconnection and use. Passive oxygen analyzers have the great advantage of stability during use. Drift is negligible over relatively long periods (at least one day). Active instruments, such as those made by the Yellow Springs Instrument Co., on the other hand, apply voltage to the probe only when switched on; probes can be disconnected during use. These instruments are subject to relatively rapid drift and may require frequent recalibration. The cost of passive analyzers is much higher than that of active ones. Passive analyzers are recommended if available.

Whatever the type of analyzer, the probe tip covered with the thin, delicate teflon membrane is the part most vulnerable to damage and requiring the most frequent attention. The membrane may easily become punctured or work slightly loose. Neither of these two problems would normally be visible, but would only show up as problems in measurement. A punctured membrane will elicit a steady drift of the reading in one direction. A loose membrane, on the other hand, will cause erratic slow readings. Unfortunately, neither of these two conditions can be corrected without replacing the membrane. Membrane replacement and subsequent stabilization and recalibration will take at least one hour. Oxygen analyzer probes differ greatly in probe design and membrane protection; in many, the probe tip is well protected by the stirrer assembly, but in others protection is minimal. Membranes can become loose merely from the transfer of the probe from bottle to bottle. In all cases, great care in the use of the probe is advisable. A probe with an unprotected tip should never be used *in situ* in tidal pools. In rare cases, probes may become chemically poisoned or covered with growths of bacteria or algae. Cleaning procedures are specialized and are covered in instrument manuals. Probes stored in water should have a little formaldehyde added to the water to stop microorganismal acitivty.

All types of oxygen analyzer require reasonably frequent calibration. This is relatively easily accomplished using fresh water saturated with oxygen. Tables are provided with each instrument which give saturation levels for fresh water at all temperatures and pressures. In practice a sample of saturated water is prepared by vigorously pouring water, with obvious churning, between two containers about 20 times. The instrument is then set to the value shown in the table of saturation for the appropriate temperature and pressure (or elevation). Calibration must not be done if the water used is changing in temperature. A quick and rough check of calibration can be done by exposing the sensor to the air. Since the oxygen content of air is fairly stable, a calibration check is possible by reference to the tables. This method should not be used to set the instrument, but is useful in detecting calibration problems in the field. It is also possible to carry out accurate calibration using water of any oxygen content, providing an accurate Winkler titration can be made.

Planning

Timing. If the experiment is to be run *in situ*, tide tables must be consulted in advance to make sure that the pool will be exposed for a long enough period (at least 2 h). The experiment should be done either early in the day or *immediately* after isolation by the tide. On dull days, up to three hours incubation may be needed. In bright weather, 1–2 h are usually sufficient. In *any* experiment, the appearance of bubbles in the light bottle is a signal to cease incubation. The time of exposure of pools can be calculated if the height of the pool above chart datum is known by use of methods for calculating the time of the tide for a specific height given in standard tide tables. Lacking this knowledge, times can be approximated by observation of the isolation time on the *previous* day and adding one hour.

Bottle preparation. B.O.D. bottles for use in incubation must be clean and dry. A washing method that ensures the removal of all traces of organic material *must* be employed. The most reliable methods are chromic acid washing, with at least two tap water and two distilled water rinses, or at least overnight soaking in a commerical decontamination cleaner such as Decon (British Drug Houses). A substitute for chromic acid may be made as follows. Dissolve 150 g of $K_2Cr_2O_7$ in minimal water (circa 200 mL) then add 4000 mL of 90% H_2SO_4. This mixture ('chromerge') is EXTREMELY caustic and it should only be handled in a fume hood and with protective gloves. Materials for spill cleanup should always be close at hand. The advantage of Chromerge lies only in its very rapid action; if fast turnaround of bottles is not needed, safer methods are advised.

General. If dry weights of samples are to be determined, a forced-air drying oven will be required. A still-air oven will not be satisfactory for material with a high water content.

Experimental design

The standard experiment. The simplest experiment would involve the investigation of photosynthetic and respiratory rates for, for example, the dominant algae in the tidal pool. This involves the incubation of similar samples in a pair of light and dark bottles. Results are applicable to whatever sample was enclosed for the time at which the test was carried out.

Comparison of species. Any number of species from the same pool can be compared by setting up simultaneous tests for each. If samples are collected and placed in paired bottles that are filled at the same time, only one initial oxygen reading is needed for the whole set. Care should be taken to ensure that samples are comparable (e.g., all whole plants or all plant apices).

Comparison of plant parts. Various parts of the same plant species can be tested together. For example, apices, mid-sections and basal portions could be compared, as could vegetative and reproductive organs.

Comparison of the same species from various habitats. Species that occur in pools are in some cases found on the open shore or even free-floating in coastal water. These can be compared to those from the pool. Similarly, other species that do not occur in pools, but may be washed in, can be compared to typical pool species.

Comparison of photosynthetic rates at various salinities. Tidal pools, especially shallow ones near to the top of the shore, may undergo rapid and large salinity changes. The effect on photosynthetic and respiratory processes can be simply modeled by incubation of pool algae in pool water diluted with various amounts of fresh water. In this case, care must be taken to determine the initial characteristics of the 'new' water, in respect to salinity and dissolved oxygen.

Plants presenting problems. Encrusting species are difficult to accommodate in the standard experiment. With care, however, rock fragments can be removed with attached plants and enclosed in the bottle. In such cases it is often appropriate to compare species on a basis of surface area instead of weight. Care should be taken with tubular species such as the green algae *Enteromorpha intestinalis* so that bubbles contained within the thallus are not transferred to the bottles. Such bubbles could interfere with dissolved oxygen levels and should be gently squeezed from the plant before incubation.

Epiphytes on living and dead animals. Many pool animals such as the Littorinid periwinkle snails, may carry significant, but invisible, epiphytes on the shell. Samples of such species can be introduced into pairs of bottles to investigate this phenomenon. If desired, controls with epiphytes removed can be prepared by wiping samples of snails with alcohol. In these cases, respiration results will be dominated by the animals but production, if present, should be obvious.

General. Many other aspects of tidal pool ecology will suggest themselves for experiment. It is important to think each one out in advance to ensure that the results will answer the question posed. It is also important to run *both* light and dark bottles in all situations. One cannot safely assume the absence of any component of photosynthesis or respiration and only the full experiment will separate the three parameters.

Field experiments. Field experiments are most easily carried out in small marine tidal or freshwater pools. In the simplest case, a spot in the pool, usually over the deepest area, is selected. Bottles are filled by lowering the B.O.D. bottle held at about 45° into the pool until the lower edge of the mouth is a few mm below the surface. Water must run smoothly into the bottle without bubbling. As the bottle nears full it is tilted slowly to the vertical to avoid trapping a bubble in the shoulder. If bubbling through the neck occurs or if a bubble is trapped, the sample should be discarded *to the side of the pool* and retaken. The second method, essential in very shallow or stratified pools, involves siphoning samples from the pool. As pools are always on a slope, this method is often superior. The siphon should be fixed in a typical spot in the pool by carefully weighing it down, i.e., with a stone. Water can be drawn through the tube, which is inserted, in turn, into each bottle. Reasonable speed in setting up the experiment is desirable so that changes in bottles before the initial is read are kept to a minimum.

To express results per square meter of pool bottom, first determine mean biomass from a standard quadrat sampling of the pool, then multiply the three results obtained from the equations given earlier by the mean biomass in g m^{-2} for the species in question. Note that weights should all be in the same units, either wet or dry.

Troubleshooting

Troubleshooting is fairly straightforward. Experience has shown that the main aspects to watch are as follows:

1. Make sure oxygen levels are well below saturation when experiments begin.
2. Terminate experiments if significant bubble formation is observed.
3. An increase in oxygen in the DARK bottle is biologically impossible and indicates an error. In this case, only net production can be calculated.
4. A decrease in oxygen in the light bottle IS biologically possible and merely indicates no net production. In such cases, net production should be expressed as zero, not a negative number.

5. Oxygen analyzer problems are rare, but can occur. A rapidly drifting reading is an indication of a punctured membrane. Replacement is easy but must be followed by a 30 minute recovery period and recalibration. Unstable readings often indicate a loose membrane. Membranes cannot be tightened and must be replaced.
6. It is important, as described earlier, to group bottles of probable similar temperature and oxygen levels for reading.
7. It is strongly recommended that the oxygen analyzer be kept in a central, dry area of the laboratory or in the field and have the students come to it, rather than moving it around.
8. Technicians, demonstrators or instructors should preferably operate the instrument, explaining the procedure to students. Usually only one analyzer is available and all the results depend on it, so very careful operation and maintenance is a must. Even if a spare instrument is available for emergencies, the change may well mask small results in some experiments.
9. Although most oxygen analyzers are claimed to be protected against the weather, experience has shown that they MUST be kept scrupulously clean and dry. If experiments are carried out in wet weather, protect the instrument by operating it within a sturdy plastic bag.

Suppliers of oxygen analyzers

1. Delta Analytical, 250 Marcus Boulevard, Hauppauge, N.Y. 11787.
2. Orbisphere Corporation, 5, Rue Gustave-Moynier, 1202 Geneva, Switzerland.
3. Yellow Springs Instrument Co., Yellow Springs, Ohio 45387.

The Winkler procedure for measurement of dissolved oxygen

Clinton J. Dawes

Dept. of Biology, University of South Florida, Tampa, FL
33620, USA

FOR THE STUDENT

Introduction

Winkler (1888) described a chemical procedure for the measurement of dissolved oxygen that is still the standard method. The modified Winkler procedure described here (Strickland and Parsons 1972; Dawes 1981) remains one of choice in major laboratories because of its high accuracy, simplicity, and low cost, in spite of excellent present day polarographic oxygen sensors (Gnaiger and Forstner 1983; Hitchman 1978). Both the chemical and polarographic procedures will allow accurate and rapid oxygen determinations. Usually the chemical analyses are carried out in the laboratory after "fixation" of free dissolved oxygen in B.O.D. bottles (Biological Oxygen Demand, 300 mL) in the field.

Fixation is carried out by the addition of manganous sulfate ($MnSO_4$) and alkaline potassium iodide (KI + KOH), which react as follows:

$$MnSO_4 + 2KOH \rightarrow K_2SO_4 + Mn(OH)_2$$

The reaction proceeds rapidly to equilibrium, producing the light colored, pink hydroxide that is oxidized by the dissolved oxygen:

$$2Mn(OH)_2 + O_2 \rightarrow 2MnO(OH)_2$$

The brown-colored manganous compound ($2MnO(OH)_2$) is allowed to settle in the bottle, shaken a second time and allowed to settle again. The compound is fairly stable and can be stored in a dark cool place for a

78

number of days, but is best analyzed as soon as possible. If no organics are present in the seawater, storage can also be done after acidification.

Once the precipitate is well settled, concentrated sulfuric acid is added and a manganous sulfate is formed:

$$MnO(OH)_2 + 2H_2SO_4 \rightarrow 3H_2O + Mn(SO_4)_2.$$

The sulfate compound reacts with the potassium iodide that was introduced in fixation:

$$Mn(SO_4)_2 + 2KI \rightarrow MnSO_4 + K_2SO_4 + I_2$$

The liberated iodine is chemically equivalent to the amount of dissolved oxygen in the sample and is titrated with a standardized sodium thiosulfate solution, $Na_2S_2O_3$. The basic reaction is an oxidation of thiosulfate to tetrathionate:

$$2Na_2S_2O_3 + I_2 \rightarrow Na_2S_4O_6 + 2NaI$$

Starch is added to the solution just before all the iodine is bound to aid in determination of the end point. The addition of starch will result in a blue color if iodine is present and will fade away as the thiosulfate is added.

Knowing the value of the thiosulfate in terms of liberated iodine and the volume of the sample titrated, the parts per million dissolved oxygen can be calculated.

Procedure

1. If a large water sample is used, the BOD bottle is filled by inverting and turning slightly vertical to avoid aeration of water. Normally, the experiments are carried out with BOD bottles.
2. Add 1 mL of manganous sulfate *followed* by 1 mL of alkaline iodide, *well below* the surface. Replace stoppers, displacing water at the top, without leaving any air bubbles. If bubbles occur, add more water before continuing.
3. Shake well for about 1 min by inverting bottle several times to distribute precipitate (if allowed to settle too soon, the dissolved O_2 at the top of the bottle may not be absorbed). The floc will settle to about one-half of the volume in two to five minutes.
4. Acidify the bottles with 1 mL of concentrated sulfuric acid by letting acid run down inside the neck of the bottle. Stopper bottle without displacing floc.
5. Shake well. Liberated iodine diffuses slowly so insure that it is distributed evenly by inversion before withdrawing a sample for titration. The precipitate will dissolve rapidly and the titration should be carried out

within an hour. Remove and weigh any macroscopic algae immediately following the dissolving of the precipitate.

6. Measure out 50 mL for titration (100 mL if O_2 concentration less than 0.1 mg-at O_2 L^{-1}).

7. Titrate with 0.01N thiosulfate until the iodine fades to a pale straw color. Use a 15 mL buret that is calibrated to 0.1 mL for titration. Add 5 mL of starch solution and titrate rapidly to the first disappearance of the blue color. Disregard reappearance of color as this is probably due to traces of iron salts or nitrites.

Calculations

If a 50 mL sample is used from a 300 mL BOD bottle, the following calculation can be used:

$$\text{mg-at } O_2 \text{ L}^{-1} = 0.1006 \times F \times V$$

where F is the standardization factor and V is the amount titrated. If other than a 50 mL sample is used then X aliquot is taken from Y mL bottle and the calculation used is:

$$\text{mg-at } O_2 \text{ L}^{-1} = \frac{Y}{Y - 2} \times \frac{5.00}{X} \times F \times V$$
$$\text{mg } O_2 \text{ L}^{-1} = 16.0 \times \text{mg-at } O_2 \text{ L}^{-1}$$

REFERENCES

Dawes, C. J. (1981). *Marine Botany.* John Wiley and Sons Publs, New York. 620 pp.

Gnaiger, E. & H. Forstner (Eds.) (1983). *Polarographic Oxygen Sensors.* Springer Verlag, Berlin. 363 pp.

Hitchman, M. L. (1978). *Measurement of Dissolved Oxygen.* John Wiley and Sons and Orbisphere Laboratories, Geneva. 255 pp.

Strickland, J. D. H. & T. R. Parsons (1972). *A Practical Handbook of Seawater Analysis.* Bulletin 167, Fisheries Research Board of Canada, Ottawa. 311 pp.

Winkler, L. W. (1888). The determination of dissolved oxygen in water. *Ber. Deut. Chem. Ges.* 21, 2843–2846.

NOTES FOR INSTRUCTORS

The procedures given above, at least the initial preservation of oxygen sample (Steps 1–3), can be used in laboratory of field conditions. The samples can be stored in a cool dark place until acidification (Step 4) is carried out. Acidification and following steps should be carried out in the laboratory, although commerical kits do permit all procedures to be carried out in the field.

The chemicals that follow should be stored in dark glass-stoppered bottles, clearly labeled and dated and kept in the refrigerator. Subsamples of the two fixatives (for Step 2) should be stored in a small field kit along with an automatic pipette or dropper that will dispense 1 mL of solution into the BOD bottles for field use. A field kit can consist of the fixative bottles along with a number of clean BOD bottles all carried in a wooden case.

Materials

1. <u>Manganous sulfate solution</u>

$MnSO_4 \cdot 4H_2O$	480 g L^{-1} distilled water
or $MnSO_4 \cdot 2H_2O$	400 g L^{-1} distilled water
or $MnSO_4 \cdot H_2O$	365 g L^{-1} distilled water

2. <u>Alkaline potassium iodide solution</u>

NaOH	500 g in 500 mL distilled water
KI	300 g in 450 mL distilled water

All solutions are made to volume

3. <u>Concentrated sulfuric acid</u>

$$H_2SO_4 \ (36N)$$

4. <u>Starch indicator solution.</u> Suspend 2 g of soluble starch in 300 mL–400 mL of distilled water. Titrate with 20% NaOH and stir vigorously until the solution clears or is slightly opalescent. Allow to stand for 1–2 h. Add concentrated HCl until solution is just acid to litmus paper. Then add 2 mL glacial acetic acid, dilute to 1 L with distilled water, and store in refrigerator.

5. <u>Standard 0.01N thiosulfate solution.</u> Dissolve *exactly* 2.9000 g $Na_2S_2O_3 \cdot 5H_2O$ in 1000 mL of water. Store in a dark bottle.

6. <u>Standard 0.01N potassium bi-iodate solution.</u> This solution must be accurately prepared. Dry KIO_3 at 100 C for 1 h, then weigh out exactly 0.3567 g. Dissolve in 200 mL–300 mL distilled water in volumetric flask, warm slightly and bring to the 1,000 mL mark with distilled water. Store in dark bottle in refrigerator.

All studies should use glass stoppered BOD bottles (300 mL) that have been acid (HNO$_3$) cleaned, rinsed, and dried. With a 300 mL volume bottle 1 mL each of manganous sulfate, alkaline potassium iodide, and concentrated sulfuric acid will be in excess and thus not limiting.

Standardization

The thiosulfate solution must be standardized before use and the F factor, or correction factor, determined. To do this, fill a 300 mL BOD bottle with distilled water and add in the following sequence: 1.0 mL conc H$_2$SO$_4$, 1.0 mL alkaline iodide potassium iodide solution. Stopper and mix by inversion. Then add 1.0 mL manganese sulfate solution and mix again. Take a 50 mL aliquot and add 5 mL 0.01N standard iodate solution (use great care in pipetting the exact amount). Allow iodine liberation to proceed at room temperature for 2 to 3 min. Then titrate with the 0.01N thiosulfate solution using a 15 mL buret. Repeat this standardization at least three times to insure accuracy. To calculate for F, use the following equation with V = titrate in mL:

$$F = \frac{5.00}{V} \text{ (for 0.01N thiosulfate)}$$

8

Adaptation to high and low irradiances measured by photosynthetic rate (^{14}C)

P. J. Harrison

Dept. of Oceanography, University of British Columbia,
Vancouver, BC, Canada V6T 2B1

FOR THE STUDENT

Introduction

Algae are able to trap photosynthetically active light (400–700 nm) because they possess the primary pigment chlorophyll *a* and accessory pigments (carotenoids, phycobilins, etc.). The light energy is passed to the reaction centers of photosystems I and II, where electron transport occurs, resulting in the production of ATP and NADPH. This simple functional structure, consisting of the two photosystems and their light-harvesting components (accessory pigments) has been termed a photosynthetic unit (PSU).

In order to discriminate between changes in pigment concentration and other changes, photosynthetic rate (amount of carbon fixed per unit time) is usually normalized to the number of cells or the amount of chl *a*. Photosynthesis (P cell^{-1} or P chl a^{-1}) is usually plotted against a range of irradiances for cultures that have been previously grown at low or high irradiances. These curves are called P versus I curves and they are approximately described by a rectangular hyperbola where the plateau is equal to P_{max} and the slope of the linear portion at limiting irradiances is called α (Fig. 8.1). I_k is the irradiance at which I is just sufficient to saturate photosynthesis. For review articles see Falkowski (1980), Prézelin (1981) and Richardson *et al.* (1983). For textbook chapters, see Dring (1982, Chapter 3) and Lobban et al. (1985, Chapter 2).

Light conditions in the sea may fluctuate rapidly and many algae have

83

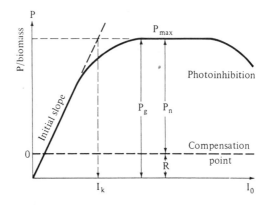

Fig. 8.1. Model light saturation curve for photosynthesis (P) versus incident irradiance (I_o), defining P_{max} = maximum photosynthesis; P_g = gross photosynthesis; P_n = net photosynthesis; R = respiration; I_k = saturating irradiance level. From Lobban et al. (1985).

adapted to respond to these fluctuations. They may change their photosynthetic capacity (number of PSU's) or vary their pigment content (increase in size of PSU). Typically, phytoplankton adapt to low irradiances by increasing their pigment content (chl *a* and accessory pigments) (Fig. 8.2). Changes in the number, shape and position of chloroplasts in the cell are of secondary importance for most species.

The photosynthetic rate is determined by measuring the amount of ^{14}C-labeled HCO^-_3, or CO_2, that is incorporated into the cells (Steemann-Nielsen 1952) or the amount of O_2 evolved. In most instances these two techniques yield approximately the same rates and in other cases the rates are different. This aspect is discussed further by Williams et al. (1983).

The objective of this laboratory exercise is to examine how algae adapt to different irradiances.

Procedure

1. Set up the filter apparatus; place a plain white 25 mm HA Millipore (0.45 μm pore size) membrane filter on top of fritted glass using forceps; clamp together. Pour the contents of the bottle into the filter funnel. Start the vacuum pump (suction should be <⅓ atm). Do not let the filter suck dry, as this slows filtration greatly and some fragile cells may lyse. Rinse the sample bottle with about 10 mL of filtered seawater. Wash the filter funnel (use a squirt bottle) with the filtered seawater and suck

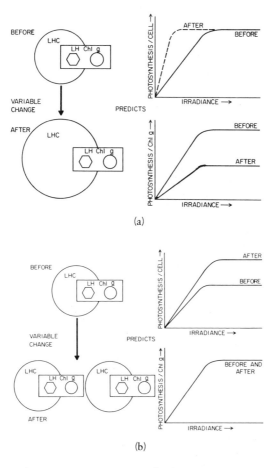

Fig. 8.2. Schematic representation of relationship between altered photosynthetic unit size (a) and photosynthetic unit density (b) and changes in the P-I curves, expressed either on a cellular basis (upper curves) or a chl *a* basis (lower curves). From Prézelin (1981).

the filter *barely* dry. (Make sure no liquid remains on the filter.) Determine the time it takes to filter 20 mL of culture. If it takes longer than 5 min, reduce the volume. Now place the volume just determined into eight bottles. Some of the light bottles have screens around them so that the irradiance reaching the cells inside will be 100, 60, 35, 21, 12, and 4% of incident light intensity. Measure the irradiance in the incubator at the bottle location. Set up two 100% light bottles and one each of the other light bottles. One of these 100% light bottles will be used as a blank (absorption correction). Set up a dark bottle (painted, or with black tape or aluminum foil around it).

2. Add a small amount (1.25 μCi) of radiolabeled $NaH^{14}CO_3$ solution to each bottle as follows. (Note the time of each addition and *stagger the additions about 3–5 min apart* so that you will be able to filter each vial after 2 h of incubation, except for one 100% light bottle (used for absorption correction), which should be filtered immediately, before significant bicarbonate uptake occurs.) Take an ampule containing 2.00 mL of suitable strength radiolabeled bicarbonate solution in 3.5% sodium chloride, score the tip with a sharp glass file and break it off. Make sure all the solution is out of the tip (tap with finger) before scoring the tip.) Carefully pipette 0.5 mL from the ampule and transfer to a ^{14}C incubation bottle. Do not pipette radioactive material by mouth. Wear surgical gloves.

3. Mix the contents of each bottle thoroughly and immediately filter the contents of one 100% light bottle (absorption correction). Put the other bottles under the lights to begin the incubation period (2 h). Make sure caps don't shade the algae. Use the same temperature and irradiance under which the cultures were grown.

4. While waiting, set up the filter apparatus with a new filter; label liquid scintillation vials and place 1 mL 10% HCl (v:v) in each. Also while you are waiting, make duplicate cell counts on the high and low light adapted cultures. Obtain the chl *a* values from the instructor (see Exp. 4).

5. At end of incubation (2 h for each sample, or note the time if not 2 h), filter off cells as above.

6. Using forceps, fold the membrane filter in half and insert it into one of the liquid scintillation vials containing 1 mL 10% HCl (to convert all the $H^{14}CO_3$ to $^{14}CO_2$, leaving only organically bound ^{14}C). The filter must lie *flat* on the bottom of the vial. After approximately 1 h add 1 mL ethyl acetate to dissolve the filter, followed by 15 mL scintillation cocktail. (Label vials on lid, *not* on the sides—why?) Store vials in a case and give them to the instructor to count on the liquid scintillation counter when you are finished.

Calculations

The uptake of carbon is calculated as follows:

$$\frac{\text{Rate of photosynthesis}}{\text{mg C m}^{-3}\text{ h}^{-1}\text{ or }\mu\text{g L}^{-1}\text{ h}^{-1}} = \frac{(R_L - R_B) \times W \times 1.05}{R \times N}$$

Where R_L and R_B = light and blank bottle counts, respectively (disintegrations per min), corrected for the counting efficiency (see as follows)

W = weight of carbonate carbon in the original sample ($mg \ C \ m^{-3}$) (see example below)

R = total activity of added bicarbonate (dpm) ($1\mu Ci =$ 2.22×10^6 dpm)

N = hrs of incubation

1.05 = factor to correct for isotopic discrimination (^{12}C vs. ^{14}C)

The instructor will give you the efficiency of the liquid scintillation counter used to measure the counts per minute (cpm) and he will explain how to convert cpm into dpm (generally dpm = cpm ÷ efficiency).

The value of W is best determined by titration or manometric analysis. (See Strickland and Parsons (1972, pp. 27 and 35). If the salinity, S‰, is known, then:

Total alkalinity = S‰ $\times$ 0.067 meq L^{-1}
Carbon alkalinity = total alkalinity $-$ 0.05
Total carbon dioxide = 0.96 $\times$ carbon alkalinity
Example: Salinity = 30.00 ‰
Total alkalinity = 2.01 meq L^{-1}
Carbon alkalinity = 1.96 meq L^{-1}
Total carbon dioxide = 1.88 meq L^{-1}
W = 22,600 mg C m^{-3}

Questions

1. Calculate the rates of photosynthesis for the high *and* low light culture using (a) $\mu g \ C \ cell^{-1} \ h^{-1}$ (b) $\mu g \ C \ \mu g \ chl \ a^{-1} \ h^{-1}$. Plot each versus irradiance of incubation. (Obtain chl *a* values from your instructor.)
2. Discuss your results, referring to the papers by Beardall and Morris (1976), Prézelin and Matlick (1980) and Jorgensen (1969), and the models presented by Prézelin (1981).
3. What are the ecological implications of adaptation to changes in irradiance by phytoplankton?

REFERENCES

Beardall, J. & I. Morris (1976). The concept of light intensity adaptation in marine phytoplankton: some experiments with *Phaeodactylum tricornutum*. *Mar. Biol.* 37, 377–387.

Dring, M. J. (1982). *The Biology of Marine Plants*. Edward Arnold, London, 199 pp.

Falkowski, P. G. (1980). Light-shade adaptation in marine phytoplankton. In, P. G. Falkowski (ed.) *Primary Productivity in the Sea.* Plenum Press, New York pp. 99–119.

Falkowski, P. G. & T. G. Owens (1980). Light-shade adaptation: two strategies in marine phytoplankton. *Plant Physiol.* 66, 592–595.

Jorgensen, E. G. (1969). The adaptation of plankton algae. IV. Light adaptation in different algae species. *Physiol. Plant.* 22, 1307–1315.

Lewis, M. R. & J. C. Smith (1983). A small volume, short-incubation-time method for measurement of photosynthesis as a function of incident irradiance. *Mar. Ecol. Prog. Ser.,* 13, 99–102.

Lobban, C. S., P. J. Harrison & M. J. Duncan (1985). *The Physiological Ecology of Seaweeds.* Cambridge University Press, Cambridge, 242 p.

Lüning, K. (1981). Light. In, C. S. Lobban & M. J. Wynne (eds.) *The Biology of Seaweeds.* University of California Press, Berkeley, pp. 326–355.

Parsons, T. R., Y. Maita & C. M. Lalli (1984). *A Manual of Chemical and Biological Methods for Seawater Analysis.* Pergamon Press, New York, pp. 173.

Prézelin, B. B. (1981). Light reactions in photosynthesis. In, T. Platt (ed.) *The Physiological Bases of Phytoplankton Ecology.* Can. Fish. Bull. No. 210, pp. 1–43.

Prézelin, B. B. & H. A. Matlick (1980). Time-course of photoadaption photosynthesis-irradiance relationship of a dinoflagellate exhibiting photosynthetic periodicity. *Mar. Biol.* 58, 85–96.

Perry, M. J., M. C. Talbot & R. S. Alberte (1981). Photoadaptation in marine phytoplankton: Response of the photosynthetic unit. *Mar. Biol.* 62, 91–101.

Richardson, K., J. Beardall & J. A. Raven (1983). Adaptation of unicellular algae to irradiance: an analysis of strategies. *New Phytol.,* 93, 157–191.

Steemann-Nielsen, E. (1952). The use of radioactive carbon (^{14}C) for measuring organic production in the sea. *J. Cons. perm. Int. Expl. Mer.,* 18, 117–140.

Strickland, J. D. H. & T. R. Parsons (1972). *A Practical Handbook of Seawater Analysis,* 2nd Edition. Fish. Res. Bd. Canada Bull. No. 167, 310 pp.

Williams, P. J. le B., K. R. Heineman, J. Marra & D. A. Purdie (1983). Comparison of ^{14}C and O_2 measurements of phytoplankton production in oligotrophic waters. *Nature* 305, 49–50.

NOTES FOR INSTRUCTORS

Materials for each set

1. filtration apparatus
2. ^{14}C bottles with light screens (16)
3. 0–1 mL pipette (adjustable) or volumetric pipette
4. Millipore filters (0.45 μm pore size, 25 mm diameter)
5. forceps
6. liquid scintillation vials (16)
7. measuring cylinder (50 mL)

Instruments

1. liquid scintillation counter
2. light meter (with quantum sensor)
3. temperature-controlled incubation with light

Reagents and chemicals

1. ampules of ^{14}C labeled HCO_3^- in 3.5% sodium chloride
2. liquid scintillation fluor (e.g. Aquasol from Fisher Products)

Cultures

Use a fast growing phytoplankter (e.g., *Thalassiosira pseudonana, Phaeodactylum tricornutum, Dunaliella tertiolecta,* etc.). Cell densities should be $\sim 10^8$ cells L^{-1}. (Refer to Algal Cultures.) This laboratory may be run with macrophytes and some modifications (see Kremer 1978 and Experiment 12).

Medium

Use f/2 (Guillard & Ryther 1962), ES/2 (Harrison *et al.* 1980), or PES (see Exp. 28) for all nutrients, trace metals and vitamins.

Light

The high- and low-light cultures should receive $>100 \mu E\ m^{-2}\ s^{-1}$ and $< 40 \mu E\ m^{-2}\ s^{-1}$ respectively. Because the high-light culture will grow faster than the low-light culture, the former must be transferred more frequently; both cultures *must* be kept in logarithmic phase growth (this can be monitored by using *in vivo* fluorescence—Exp. 4B). The adaptation time should be 5–10 generations. Since the low-light cultures will grow at about half the rate of the high-light cultures, their adaptation period will be at least twice as long.

Chl a

The students will not have time to determine chl *a*; this can be quickly estimated from *in vivo* fluorescence, provided a fluorometer that has been previously calibrated is used (see Experiment 4).

Incubation bottles

Use borosilicate (e.g., Pyrex) screw cap bottles (60 mL) and wrap nylon window screening (obtain from hardware store) around them. Use tape or glue the screening. The following is a guide: 1 layer of screen = 60% transmission; 2 layers = 35%; 3 layers = 21%; 4 layers = 12% and 6 layers = 4%, but you should check this with a light meter. Label the bottles clearly.

Filter apparatus

A vacuum pump with a gauge to set the amount of vacuum is required. Make sure you have a flask (trap) between the vacuum pump and the main vacuum flask to catch any overflow (and protect the pump) in case the students forget to empty the vacuum flask when it gets full.

Isotopes

Students should be instructed in the proper and safe use and disposal of isotopes. Consult with the radiation officer in your institution if you are not certain.

Incubation time

Two hours was chosen because this is a standard incubation period. If the lab must end after 3 h, then 1-h incubation is recommended. To insure enough counts, especially on the low-light incubations, the amount of isotope that is added should be doubled. Depending on the size of the class, it may also be possible to have the students come in early and set up the experiment and inoculate the bottles and then return 2 h later. (Note: It takes about 4 h to complete the lab if a 2-h incubation is used.) Another possibility is to use a 24-h incubation, but bottle effects become more pronounced under these circumstances.

Dark bottle

There is considerable debate on whether a dark bottle is a suitable correction for bacterial effects. A blank bottle (corrects for filter absorption, etc.) is frequently used, but this can be viewed as a minimal correction. The students should be encouraged to calculate photosynthetic rates for low-light bottles using both blank and dark bottle corrections to see the difference for themselves.

Filtration

The students should be made aware of the *importance* of this *simple* procedure. Many flagellates and dinoflagellates are extremely fragile and cell lysis can *easily* occur.

Photosynthetic rates glassware

If the class is small, then an O_2 electrode may be used in place of or in addition to ^{14}C to measure photosynthetic rates (see Experiment 7). Because this lab requires special glassware, it is not practical to work individually. Students may work in groups (two or three) or the labs may be arranged so one pair does this lab while other pairs are doing other labs, rotating each week. The requirement for incubation bottles may be reduced by one pair doing the high-light culture and another pair doing the low-light culture.

Comments

The instructor should explain the theory of the ^{14}C technique and the problems that are still associated with the method (see reviews by Carpenter & Lively 1980; Peterson 1980). The students should be introduced to gross and net photosynthesis and why, in these experiments (2 h incubation), the technique is measuring in between gross and net photosynthesis, but probably closer to net.

Freshwater phytoplankton can be used in place of marine phytoplankton with a few modifications. Use BOD (biological oxygen demand) bottles or any screw-cap bottle. Fill the bottles full of culture medium; inject the ^{14}C solution into the bottom of the incubation bottle and screw cap firmly, making sure there is no air space in the bottle. Because of the lower pH in freshwater than seawater, some of the $H^{14}CO_3^-$ may be converted to $^{14}CO_2$ and may be lost in the air space and not, consequently, available for uptake during the incubation period.

Additional References

Carpenter, E. J. & J. S. Lively (1980). Review of estimates of algal growth using ^{14}C tracer techniques. In, P. G. Falkowski (ed.) *Primary Productivity in the Sea*. Plenum Press, New York, 161–178.

Cooper, T. G. (1977). *The Tools of Biochemistry*. John Wiley & Sons, New York, 423 pp.

Guillard, R. R. L. & J. H. Ryther (1962). Studies on marine planktonic diatoms, I. *Cyclotella nana*, Hustedt and *Detonula confervacea* (Cleve) Gran. *Can. J. Microbiol.* 8, 220–239.

Harrison, P. J., R. E. Waters & F. J. R. Taylor (1980). A broad spectrum artificial seawater medium for coastal and open ocean phytoplankton. *J. Phycol.* 16, 28–35.

Iverson, R. L., H. F. Bittaker & V. B. Myers (1976). Loss of radiocarbon in direct use of Aquasol for liquid scintillation counting in solutions containing ^{14}C- Na HCO$_3$. *Limnol. Oceanogr.*, 21, 756–758.

Kremer, B. P. (1978). Determination of photosynthetic rates and ^{14}C photoassimilatory products of brown seaweeds. In, J. A. Hellebust & J. S. Craigie (eds.) *Handbook of Phycological Methods: Physiological and Biochemical Methods.* Cambridge University Press, New York, pp. 270–283.

Lean, D. R. S. & B. K. Burnison (1979). An evaluation of errors in the ^{14}C method of primary production measurement. *Limnol. Oceanogr.*, 24, 917–928.

Peng, C. T. (1981). *Sample Preparation and Liquid Scintillation Counting.* Amersham Corp., Illinois, 112 pp.

Peterson, B. J. (1980). Aquatic primary productivity in the ^{14}C-CO$_2$ method: A history of the productivity. *Ann. Rev. Ecol. Syst.*, 11, 359–385.

9

Qualitative analysis of pigments

David J. Chapman

Dept. of Biology, University of California, Los Angeles, CA
90024, USA

FOR THE STUDENT

Introduction

The chlorophylls and carotenoids are the essential pigments for photosynthesis, involved in light energy harvesting and, in the case of some carotenoids, photoprotection. The chlorophylls and carotenoids of algae are also very important as taxonomic criteria for individual classes and divisions, and as phylogenetic markers. The significance of this statement is easily understood when one remembers that many divisions are "color-coded" (e.g., red algae (Rhodophyta), yellow-green algae (Xanthophyta), brown algae (Phaeophyta), golden-brown algae (Pyrrophyta), etc.). You are referred to a number of review articles (Goodwin 1980; Jeffrey 1976, 1980; Larkum and Barrett 1984; Liaaen-Jensen 1978, 1985) that discuss pigment distribution.

Visual observation is inadequate in determining the particular combination of chlorophylls and carotenoids, due to masking, *in vivo* spectral characteristics that are different to those *in vitro*, quantitative variations and the fact that nearly all carotenoids are "yellow-orange" and chlorophylls are "green."

Simple, effective techniques of pigment separation are important. This exercise demonstrates such techniques and illustrates the major chlorophyll/carotenoid differences between classes (Table 9.1). In Experiment 11, you isolate and separate the proteinaceous pigments, the phycobiliproteins that are found additionally in four groups; Cyanobacteria, Rhodophyta, Cryptophyta and Glaucophyta. Those pigments are separated by

93

Table 9.1. *Principal chlorophylls and carotenoids of algal classes*[a,b]

CYANOBACTERIA
  Chlorophyll *a*
  β-carotene, echinenone, zeaxanthin
  myxoxanthophyll
RHODOPHYTA
  Chlorophyll *a*
  β-carotene, α-carotene, lutein zeaxanthin
EUGLENOPHYCEAE
  Chlorophyll *a, b*
  β-carotene, zeaxanthin, diadinoxanthin, neoxanthin
DINOPHYCEAE
  Chlorophyll *a*, c_2
  β-carotene, dinoxanthin, peridinin, diadinoxanthin, pyrroxanthin
  Note: In those dinoflagellates with a "Chrysophytan-like" chloroplast, e.g.,
  Gymnodinium veneficum, Peridinium balticum, Peridinium foliaceum, and
  Gyrodinium aureolum, peridinin is replaced by fucoxanthin and sometimes
  19[1]-hexanoyloxyfucoxanthin and chlorophyll c_1 are present.
BACILLARIOPHYCEAE
  Chlorophyll *a*, c_1, c_2
  β-carotene, diatoxanthin, diadinoxanthin, fucoxanthin
CHRYSOPHYCEAE
  Chlorophyll *a*, c_1, c_2
  β-carotene, zeaxanthin, violaxanthin, neoxanthin, fucoxanthin
PRYMNESIOPHYCEAE
  Chlorophyll *a*, c_1, c_2
  β-carotene, diatoxanthin, diadinoxanthin, fucoxanthin
  Fucoxanthin replaced by 19[1]-hexanoyloxyfucoxanthin in *Emiliania huxleyi.*
CRYPTOPHYCEAE
  Chlorophyll *a*, c_2
  α-carotene, crocoxanthin, monadoxanthin, alloxanthin
CHLOROPHYCEAE-CHAROPHYCEAE-ULVOPHYCEAE
  Chlorophyll *a, b*
  β-carotene, lutein, zeaxanthin, violaxanthin, neoxanthin
  Siphonaxanthin and its ester siphonein in the Ulvophyceae
PHAEOPHYCEAE
  Chlorophyll *a*, c_1, c_2
  β-carotene, violaxanthin, fucoxanthin
XANTHOPHYCEAE
  Chlorophyll *a*, c_1, c_2
  β-carotene, diatoxanthin, diadinoxanthin, heteroxanthin, vaucheriaxanthin,
  neoxanthin,
  Chlorophylls c_1, c_2 present in very low amounts

(*Continued*)

Table 9.1. (*Continued*)

EUSTIGMATOPHYCEAE
  Chlorophyll *a*
  β-carotene, vaucheriaxanthin, neoxanthin
CHLOROMONADOPHYCEAE
  Chlorophylls, *a*, c_1, c_2
  β-carotene, dinoxanthin, diadinoxanthin, heteroxanthin, vaucheriaxanthin

---

a) For the distribution of trace carotenoids or carotenoids of limited distribution refer to review literature listed.

b) The order of separation can be checked from data in Davies (1976), but the list of carotenoids is an approximate order of decreasing R_f. As a general guide, note that α, β-carotenes develop close to the solvent front, and may not separate. Chlorophyll *a*, with chlorophyll *b* trailing, develop behind the carotenes but in front of most of the xanthophylls. Fucoxanthin, peridinin, neoxanthin and myxoxanthophyll develop behind most of the other xanthophylls, but in front of the chlorophyll c_1 c_2, which are usually the slowest developing pigments, with lowest R_f.

techniques for water soluble compounds of high molecular weight such as electrophoresis using aqueous buffers. In this experiment, you employ adsorption chromatography with organic solvents, which is a procedure used typically for compounds soluble in organic solvents, such as lipids, chlorophylls and carotenoids. The particular procedure you will use is Thin Layer Chromatography (TLC). It is quick, very effective, and repeatable.

Procedure

Precautions. Chlorophylls and carotenoids are unstable compounds. If they are not handled carefully they will undergo chemical changes that can make separations and identifications more difficult. They are very sensitive to exposure to light and air, undergoing isomerisations and oxidations that change the colors and chemical structure. Photochemical alterations are accelerated when pigments are adsorbed on chromatographic surfaces. The presence of even a small amount of acid in solvents will remove the magnesium from chlorophyll to produce phaeophytins and cause *cis/trans* isomerisation of carotenoids. Strong alkali will cleave the phytol from chlorophylls to yield chlorophyllides and can produce chemical alteration in certain carotenoids (e.g., fucoxanthin, peridinin, and siphonein, the fatty acid ester of siphonaxanthin).

Pigments must be handled carefully with minimum exposure to light and air. Whenever possible, work in dim light; keep solutions in glassware covered with black cloth (or, if possible, stored in red actinic glassware). Chromatographic chambers must be covered with black cloth. If solutions or evaporated samples are to be stored for longer than an hour, they should be flushed with nitrogen and stored in the refrigerator or freezer. If you need to keep your pigment samples for more than a day (particularly chlorophylls or total extracts) store them evaporated, under N_2 in a freezer. With care they can last, with only minor changes for about 4 weeks.

Whenever possible, work in a functioning fume hood, and *NO SMOKING!* When handling extractions and solvents, use gloves if possible. Use freshly opened bottles of solvents. This is particularly important for methanol, ethanol and acetone, which readily take up water from the atmosphere. An opened capped can of diethyl ether should be stored in an explosion-proof refrigerator, and kept for no longer than a month because of the possibility of the formation of explosive peroxides.

Extraction and chromatography. You will be provided with cultures of unicells and samples of freshly collected macroalgae (or material that has been previously harvested and stored at $-20°C$).

1. Harvest the unicells by centrifugation and wash twice at the centrifuge with distilled water. Macroalgae samples should be quickly rinsed with distilled water and blotted damp-dry with filter paper.
2. Grind the algal mass of unicellular cultures and extract with methanol and approximately 5–10 mg $MgCO_3$ in a Ten-Broeck hand homogenizer. Cut the macroscopic thallus algae (brown, green) into small pieces and extract in methanol with 5–10 mg $MgCO_3$ in a chilled pestle and mortar with sand or grinding alumina. Extraction of macroalgae can be facilitated by pre-freezing cut pieces in liquid N_2.

    The volume of methanol per extraction will usually be about 5–10 mL for the unicells and 15–20 mL for the macroalgae.
3. Centrifuge the extracts in a clinical centrifuge at full speed for 5–10 minutes. Recover the colored supernatant, which should be clear, and re-extract the pellet as in step 2. Two or three extractions are usually sufficient to extract the bulk of the pigments.
4. Combine the supernatants (approximately 40–60 mL from each algal extract) and add to a 250 mL separatory funnel. Add approximately 30 mL of diethyl ether to the pigment solution and gently tilt the funnel to produce a homogenous solution.

    CAUTION: FUME HOOD AND NO SMOKING

You should be working in dim light and, if possible, you should wrap the separatory funnel in black cloth during these phasing operations.

5. Add cold 5% NaCl solution carefully to the separatory funnel by pouring down the side (add approximately 2 volumes of NaCl relative to the total methanol-diethyl ether solution). The solution in the funnel will separate into 2 layers, an upper, epiphase of diethyl ether containing most of the pigments and a lower hypophase containing the methanol-NaCl with occasionally some residual pigment. Stopper the funnel and *gently* tilt it back and forth a number of times. Hold the funnel upside down (hand on stopper!) and open the stopcock gently to release pressure that builds up. Don't point the spout of the funnel at anyone. (Diethyl ether boils at 37°C.) If the hypophase is colorless, discard from the separatory funnel, and proceed to Step 7.

6. If the hypophase is still pigmented, it must be washed. Drain the hypophase and epiphase separately from the funnel, and add the hypophase back to the funnel. Add another 30 mL diethyl ether, and repeat the shaking process. Discard the hypophase which should now be colorless.

7. Combine the diethyl ether extracts in the funnel and wash the diethyl ether with cold water to remove any residual methanol, and discard the aqueous hypophase. Repeat the cold water wash once.

   Note: a) diethyl ether is slightly soluble in water, so the volume will decrease with washing. If the volume of diethyl ether diminishes to too small a volume with which to conveniently work, simply add more diethyl ether; b) phasing the pigments into diethyl ether must be done *gently*. Shaking, etc. will frequently cause emulsions. These can be broken by addition of ethanol, methanol or more sodium chloride.

8. Recover the diethyl ether extract from the funnel and dry the solution by slow passage through a 1 cm layer of granular anhydrous sodium sulfate in a medium porosity sinter glass filter (size 60). Use a buchner funnel and slight vacuum. Wash the sodium sulfate with a few milliliters of diethyl ether to remove any adsorbed pigment. Evaporate the diethyl ether to dryness in a rotary evaporator under vacuum. If this is not available, small aliquots (e.g., 15 mL) can be very quickly evaporated to dryness under a stream of N_2 in a fume hood. The dried pigments will form a shiny surface in the round-bottom flask or beaker. If you used a rotary evaporator do not discard the diethyl ether in the recovery vessel down the sink, but place in an appropriate organic waste container.

9. Chromatography: Since the various chlorophylls and carotenoids demonstrate a wide array of polarity and adsorption characteristics, no one chromatographic system will separate all the pigments. Accordingly, you will chromatograph each extract on two systems:

a) Silica gel with 25% acetone in petroleum ether
b) Cellulose with 2% *n*-propanol in petroleum ether

50 minutes before use, line the chromatographic chamber with filter paper (Whatman #1) and add 120 mL solvent per chamber, pouring the solvents down the side to moisten the filter paper. This will ensure vapor saturation. Seal the lid to the chamber with a layer of vacuum grease. Dissolve the evaporated pigment extract in a few milliliters of acetone to produce a concentrated solution and apply to both TLC plates about 20 mm from the bottom as 15–20 mm bands (see Figure 9.1). You should apply three or four different extracts to each plate (same set to both cellulose and silica gel). If you are chromatographing only one extract for preparative chromatography (see section that follows on preparative chromatography), let the band be 150 mm in width.

10. Apply the pigment extracts by means of drawn pasteur pipettes (Figure 9.2). Draw up about 0.5 mL into the pipette with a rubber bulb. Outflow is best controlled by holding a finger over the wide end. Caution: Do not scratch the surface of the adsorbent. Scratches interfere with the separation.

11. When the appropriate extracts have been applied, dip the plate in a trough of acetone (5 mm depth) in a separate chamber and allow to develop for about 10 mm to form a thin sharp band. This will ensure that the pigments develop as a sharp band all from the same origin and thus facilitate comparisons. Each band will have spread out as it was applied. (Applied extracts should be covered with dark card to shield from light.)

12. Remove the plate from the acetone, and allow to dry (about one minute). Mark the starting (origin) point with a scratch mark at the edge of the adsorbent. Place the plate in the appropriate solvent chamber

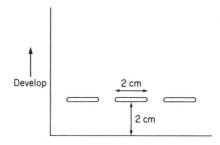

Fig. 9.1. Diagram of chromatogram at start.

Fig. 9.2. Drawn Pasteur pipette for applying pigment to chromatography plates.

and cover with a black cloth. Let the plates develop to about 25 mm from the top, remove, mark the solvent front, dry, and diagram the various carotenoids and chlorophylls and their R_f values.

$$R_f = \frac{\text{Distance pigment moved from origin}}{\text{Distance solvent moved from origin}}$$

13. Identification of pigments: This can be achieved with a good measure of reliability by noting *relative* chromatographic positions and recording the absorption spectra. R_f values, *per se*, have little meaning, since they are affected by so many experimental variables. Relative chromatographic positions of carotenoids and chlorophylls are given in Table 9.1.

Preparative chromatography

1. The small pigment bands contain insufficient pigment to obtain good absorption spectra. To record an accurate absorption spectrum it is suggested that for each alga a full width band of the extract be streaked out onto a silica gel plate and chromatographed as described above.
2. Scrape off each individual principal pigment zone (from the plate in step 1) with a razor blade (shield unscraped bands from light), and grind the adsorbent-pigment silica gel to a fine powder with a spatula in a small beaker.
3. Add the powder to a pasteur pipette that has been blocked with glass wool. Add 2–3 mL of acetone (for chlorophylls) or ethanol (for carotenoids) and recover the eluted pigments. Add sufficient solvent until the powder is white. Record the absorption spectra for chlorophylls (700 nm to 400 nm) and carotenoids 600 nm to 350 nm). Compare with published values (e.g., Davies 1976; Holden 1976; Jensen 1978). (See also Table 12.1).

Note: Chlorophylls appear green on the plates, carotenoids appear yellow or orange.

Questions

1. Identify the principal carotenoids and the chlorophylls present. Compare with the published literature.
2. Can you identify on your chromatograms alteration products of chlorophyll (e.g., phaeophytins, chlorophyllides) and *cis* isomers of the carotenoids?
3. How would you determine if *cis* isomers of carotenoids, for example, are real products or artifacts of extraction?
4. Your absorption maxima may differ from published ones by 2–3 nm. What is your explanation for this?
5. Other, colorless compounds will be present on the chromatogram. What classes of compounds do you think these are? (The location of many may be visualized by illuminating the plate with long wavelength UV light in a dark room.) CAUTION: Don't look directly at the UV lamp when it is on.
6. Why don't you use old stationary phase cultures?

REFERENCES

Davies, B. H. (1976). Carotenoids. In, T. W. Goodwin (ed.) *Chemistry and Biochemistry of Plant Pigments, Vol. II*. Academic Press, New York, pp. 38–165.

Goodwin, T. W. (1980). *The Biochemistry of the Carotenoids*. 2nd Ed. Volume I. *Plants*. Chapman and Hall, London, pp. 1–377.

Holden, M. (1976). Chlorophylls. In, T. W. Goodwin (ed.) *Chemistry and Biochemistry of Plant Pigments, Vol. II*. Academic Press, New York, pp. 2–37.

Jeffrey, S. W. (1976). The occurrence of chlorophyll c_1 and c_2 in algae. *J. Phycol.* 12, 349–354.

Jeffrey, S. W. (1980) Algal pigment systems. In, P. G. Falkowski (ed.) *Primary Productivity in the Sea*. Plenum Press, New York, pp. 33–58.

Jensen, A. (1978). Chlorophylls and carotenoids. In, J. A. Hellebust & J. S. Craigie (eds.) *Handbook of Phycological Methods. Physiological and biochemical methods*. Cambridge University Press, Cambridge, pp. 59–70.

Larkum, A. W. D. & J. Barrett (1983). Light harvesting processes in algae. *Adv. Bot. Res.* 10, 3–222.

Liaaen-Jensen, S. (1978). Marine carotenoids. In, P. Scheuer, (ed.) *Marine Natural Products, Vol. 2*. Academic Press, New York, pp. 2–75.

Liaaen-Jensen, S. (1985). Carotenoids of lower plants – recent progress. *Pure Appl. Chem. 57*, 649–658.

NOTES FOR INSTRUCTORS
Algae

A representative from each of a number of classes is recommended.

*Euglenophyceae**	
e.g., *Euglena gracilis* Z strain	UTEX 753
	Carolina 15-2802
Dinophyceae	
e.g., *Amphidinium carterae*	UTEX 1002
	Carolina 15-3240
*Bacillariophyceae**	
e.g., *Phaeodactylum tricornutum*	UTEX 642
	Carolina 15-3065
*Chrysophyceae**	
e.g., *Ochromonas danica*	UTEX 1298
	Carolina 15-3200
*Prymnesiophyceae**	
e.g., *Prymnesium parvum*	UTEX 995
*Cryptophyceae**	
e.g., *Rhodomonas sp.*	UTEX 2163
*Chlorophyceae**	
e.g., *Spongiochloris*	UTEX 1238
*Charophyceae**	
e.g., *Klebsormidium*	UTEX 322
*Ulvophyceae**	
Any natural population, e.g., *Codium, Cladophora, Ulva*	
*Phaeophyceae**	
Any natural population or *Ectocarpus siliculosus*	UTEX 2008
Xanthophyceae	
e.g., *Vaucheria sessilis*	UTEX 146
Eustigmatophyceae	
e.g., *Pleurochloris commutata*	UTEX 310
*Cyanobacteria**	
e.g., *Nostoc muscorum*	UTEX 486

Classes marked with an asterisk are the best examples.

Rhodophyceae are not satisfactory material. The amount of chlorophyll or carotenoids per unit mass is frequently much lower than in other classes.

Cell mass from approximately 2 liters of culture is usually sufficient to provide enough for 12 groups. Use cultures that are still actively growing or early stationary phase.

You can save time if the cultures are harvested (Step 1) before the beginning of class and kept frozen at $-10°C$ prior to pigment extraction.

Supplies

Supplies for each pair of students. Each pair extracting one alga.

1. pestle and mortar (small) or Ten Broek homogenizer (40 mL) (1)
2. stoppered erlenmeyer flasks (50 mL) (3)
3. stoppered erlenmeyer flasks (25 mL) (3)
4. 125 mL buchner flask with side arm (1)
5. appropriate centrifuge tubes (15 mL) (6)
6. 30 mm medium porosity sintered glass filter with drilled stopper to fit side arm buchner (1).
7. separatory funnel (250 mL) with teflon stopcock (1)
8. ring stand (1)
9. 250 mL round bottom flask for rotary evaporator (1)
10. 50 mL beakers (6)
11. 25 mL beakers (6)
12. prelaid 20 × 20 cm silica gel TLC plates (commercially available from supply houses) (2). At the beginning of the class activate the plates by heating for 1 hour at 110°C. (Store in desiccator or desiccator box if not used immediately.)
13. prelaid 20 × 20 cm cellulose plates (2)
14. standard-sized chromatographic chambers (2). Two chambers per two or three pairs is sufficient. Each chamber can hold four to five plates.
15. spotting pipettes (6). These are prepared by pulling Pasteur pipettes as illustrated in Figure 9.1.
16. spatulas
17. Whatman #1 filter paper (sheets)
18. single-edged razor blades
19. Pasteur pipettes
20. black cloth

Chemicals needed – all reagent grade. These can be supplied as a class supply.

1. Acetone*
2. Methanol*
3. Diethyl ether, anhydrous* (HAZARDOUS SOLVENT, CAUTION.)
4. n-Propanol
5. Petroleum ether (30–60° or 45–60° B.P.)
6. Anhydrous sodium sulfate, granular
7. Magnesium carbonate
8. Grinding alumina or fine grinding sand
9. N_2 cylinder
10. Disposable gloves.

*fresh bottles. Solvents in pre-opened bottles may have taken up water from the atmosphere.

Laboratory equipment.

1. Table-top centrifuge
2. Spectrophotometer with cuvettes
3. Rotary evaporator (if available)

Scheduling

It is suggested that students work in pairs, each pair being responsible for the *chromatography* of either a set of three to four different algal classes or one algal class to provide sufficient pigment for spectral identification. Each pair is also responsible for the extraction and work-up of one species (Steps 1–8). If a full three-hour laboratory period is not available, the instructor and laboratory assistant can prepare the extracts prior to the lab, usually the day before. The best place to stop is when the pigment extracts in ether have been evaporated to dryness (Step 8). These may be flushed with nitrogen, stoppered and stored in the freezer until the next laboratory.

The activation of the thin layer plates should be commenced at the beginning of the lab period, to ensure they are ready for the students.

10

Formation and analysis of secondary carotenoids

David J. Chapman

Dept. of Biology, University of California, Los Angeles, CA
90024, USA

FOR THE STUDENT

Introduction

A number of green algae, particularly the coccoid forms in the order Chlorococcales, deposit carotenoids outside the chloroplast under conditions of nutrient deficiency. These "extra-plastidic" or secondary carotenoids are typically keto-carotenoids, different to those found in normal chloroplasts (Dersch 1960; Czygan 1968; Goodwin 1980; McLean 1967).

Although there is a body of information (Kessler & Czygan 1967; Goodwin 1971) on which algae form which carotenoids, very little is known about the biosynthesis and deposition, or why the algae form these carotenoids, and why they are frequently different to the normal carotenoids of the chloroplast.

Localization of these carotenoids is not uniform, varying from species to species (Mayer & Czygan 1969). Although deficiencies of different nutrients will cause the formation of these carotenoids, nitrogen deficiency is the most commonly observed cause.

This laboratory experiment will examine this phenomenon. You will experimentally induce formation of these carotenoids, identify them and also determine if there is a significant quantitative change or degradation of the normal chloroplast carotenoids and chlorophyll. This work incorporates the cell-counting technique from Experiment 1, chlorophyll analysis from Experiment 4, and techniques for pigment separation from Experiment 9.

Procedure

In this experiment, you will be working with organic solvents and light-sensitive pigments. The precautions outlined in the procedure in Experiment 9 are applicable here and should be carefully read and followed.

Since this is a quantitative experiment, it is very important that you keep accurate records of all weights, cell numbers, volumes, and dilutions.

This experiment is in two parts. The first part involves the pigment analysis of normal cells, grown under nutrient-sufficient conditions. The second part, carried out four to six weeks later, involves the pigment analysis of cultures grown under nitrogen deficiency.

Part I. You will be provided with 1L of culture of a coccoid green alga. The cells from a similar 1 L culture will have been transferred to 1L of nitrogen-free medium.

1. Take a small aliquot and perform a cell count, as described in Experiment 1. This will allow you to calculate the number of cells that will be extracted for chlorophyll and carotenoid analysis.
2. Examine the cells under the microscope, paying particular attention to the shape, size, location, color, and number of chloroplasts.
3. Divide the approximately 1L of culture into two aliquots of accurately known volume; one ca. 200 mL, the other ca. 800 mL.
4. Take the 200 mL aliquot. This will be used for the determination of chlorophylls *a* and *b*. Recover the cells by centrifugation (10 min at 10,000 $\times$ g). Wash the cells twice at the centrifuge with distilled water and determine the amounts of chlorophylls *a* and *b* according to the procedure outlined in Experiment 4.

   Express your results as mg chlorophyll *a* and *b* per 10^6 cells.
5. You will use the remaining ca. 800 mL aliquot for carotenoid analyses. You will not attempt to quantify every carotenoid, but only β-carotene, and the total principal xanthophylls (lutein, zeaxanthin, violaxanthin and neoxanthin). Recover the cells by centrifugation (10 min at 10,000 $\times$ g), wash at the centrifuge with distilled water as above. Extract the cells and prepare the extracts for chromatographic separation according to the procedures given in steps 2–8 of Experiment 9.
6. Dissolve the evaporated pigment extract in 3 mL of acetone. Make sure all the pigments are dissolved. Transfer 2 mL of this pigment to a 5 mL beaker. This will be used for chromatography. Allow the remaining 1 mL of extract to dry, flush the round-bottom flask with

N_2, stopper and store in the freezer. This extract will be used for comparative purposes with the nutrient-deficient cultures.

7. Apply the 2 mL pigment extract as a 150 mm-wide band to an activated silica gel plate according to instructions of Steps 9 and 10 in Experiment 9.

8. Chromatograph the extract on the plate by the procedure outlined in Steps 11 and 12 of Experiment 9. Use a solvent system of 25% acetone in petroleum ether (40–60° b.p.).

   The carotenoids should develop in the following sequence, from the solvent front: β-carotene: chlorophyll *a*, chlorophyll *b*: lutein: zeaxanthin: violaxanthin: neoxanthin.

9. When chromatography is complete, remove the plate from the chamber and allow to dry (ca. 1 minute).

10. Recover the principal pigment zones by scraping from the plate with a single-edged razor blade (as in Experiment 9).

11. Combine the zeaxanthin, lutein, violaxanthin, and neoxanthin zones. Keep the β-carotene band separate. Elute the β-carotene from the silica gel powder with ethanol and separately elute the xanthophylls from the combined powder also with ethanol. Follow the instructions for preparative chromatography in Experiment 9, except that you will use a 15 mL, medium porosity sinter glass funnel instead of the pasteur pipette. Recover the eluate by vacuum suction into a 25 mL sidearm buchner flask. Use about 10 mL of ethanol. All the pigments are eluted when the silica gel powder is white.

12. Recover separately the β-carotene and mixed xanthophylls eluates. Record the volumes and measure the absorption spectra between 550 nm and 350 nm. For β-carotene the maximum absorbance occurs at approximately 450 nm, for the mixed xanthophylls between 440 nm and 450 nm. Use a 1 cm path-length cuvette.

13. Calculate the *concentration* of your pigments in the eluate from the equations

$$C = \frac{A \times 10}{E_{1cm}^{1\%}}$$

   C = Concentration in mg mL^{-1}
   A = Measured absorbance at wavelength of maximum absorbance
   $E_{1cm}^{1\%}$ = 2600 for β-carotene
   $E_{1cm}^{1\%}$ = 2500 for xanthophyll mix

Convert your values to mg carotenoid per 10^6 cells.

Part II. You will be provided with 1 L of culture maintained for four weeks under nitrogen-deficient conditions. This culture started the four weeks nitrogen starvation with the same number of cells per liter as the culture in Part I.

14. Follow Steps 1 through 9 of the protocol. Additionally, at Step 7, streak a 5 mm band of your nutrient-sufficient pigment extract (obtained by redissolving the dried 1 mL pigment extract (Step 6, above) in 1 mL acetone), adjacent to the edge of the main pigment band.
15. A comparison of your pigments from the nutrient-deficient cells and the standard band of the pigments from the nutrient-sufficient cells will allow you to identify the additional keto-carotenoids that have been formed. These, in order of decreasing R_f, are echinenone, canthaxanthin and astaxanthin. The first two develop behind β-carotene, astaxanthin just in front of neoxanthin.
16. Recover the principal pigment zones by separating from the plate with a single-edged razor-blade.
17. Keep the β-carotene zone separate. Combine the bands with the zeaxanthin, lutein, violaxanthin and neoxanthin. Combine the zones containing the keto-carotenoids (echinenone, canthaxanthin and astaxanthin). Recover the combined pigments by elution with ethanol as in Step 11.
18. Record the absorption spectra between 550 nm (600 nm for the extraplastidic keto-carotenoids) and 350 nm (as in Step 12).
19. Calculate the *concentrations* of your pigments using the equation and procedure of Step 13. For the extra-plastidic keto-carotenoids use an extinction value of $E_{1cm}^{1\%} = 2200$.

    Depending upon the composition of the keto-carotenoids the wavelength of maximum absorbance will be between 450 and 460 nm.
20. Convert your values to mg carotenoid per 10^6 cells.

Questions

1. Why are your results expressed in terms of cell number and not dry weight?
2. What was the increase in cell numbers after transferring to nitrogen-deficient culture medium?
3. Did you observe any extraplastidic carotenoids in the nutrient-sufficient cultures?
4. What visual microscopic differences do you observe between cells of nutrient-sufficient and nutrient-deficient cultures?
5. In the nutrient-deficient cultures, did you observe any other qualitative pigment changes in addition to the appearance of echinenone, canthaxanthin and astaxanthin?
6. Do your data suggest that the nutrient-deficient cells demonstrate degradation (turnover or catabolism) of the chlorophylls and/or normal carotenoids?

7. How would you design the experiment to test the possibility that changes in pigment concentration are due to dilution effects (i.e., cells keep dividing, but pigment synthesis stops)?
8. Can you think of possible roles for these extraplastidic carotenoids?

REFERENCES

Much of the work on extraplastidic carotenoids has been published in German. Accordingly, only five references to original articles, listing organisms showing extraplastidic carotenoid formation, are provided, together with two general references.

Czygan, F.-C. (1968). Sekundär-Carotinoide in Grünalgen. I. Chemie, Vorkommen und Faktoren, welche die Bildung diese Polyene beiinflussen. *Arch. Mikrobiol.*, 61, 81–102.

Dersch, G. (1960). Mineralsalzmangel und Sekundärcarotinoiden in Grünalgen. *Flora* 149, 566–603.

Kessler, E. & F.-C. Czygan (1967). Physiologische und biochemische Beitrage zur Taxonomie der Gattungen *Ankistrodesmus* und *Scenedesmus*. I. Hydrogenase, Sekundär-Carotinoiden und Gelatine-verflüssigung. *Arch. Mikrobiol.* 55, 320–326.

McLean, R. (1967). Primary and secondary carotenoids of *Spongiochloris typica*. *Physiol. Plantarum* 20, 41–47.

Mayer, F. and F.-C. Czygan (1969). Anderungen der Ultrastrukturen in den Grünalgen *Ankistrodesmus braunii* und *Chlorella fusca* var. *rubescens* bei Stickstoffmangel. *Planta* (Berl.) 86, 175–185.

Goodwin, T. W. (1971). Algal Carotenoids. In, T. W. Goodwin (ed.) *Aspects of Terpenoid Chemistry and Biochemistry*. Academic Press, New York, pp. 315–356.

Goodwin, T. W. (1980). Distribution of Carotenoids. In, T. W. Goodwin (ed.) *Chemistry and Biochemistry of Plant Pigments*. 2nd edition. Vol. 1. Academic Press, New York, pp. 225–261.

NOTES FOR INSTRUCTORS

It is suggested that students work in groups of four: one pair for chlorophyll analysis and cell counts, one pair for carotenoids.

Supplies for each quartet of students

The supplies and equipment listed for qualitative analysis of pigments (Experiment 9) and estimation of biomass (Experiment 1) are adequate. The supplies for chlorophyll analysis depend upon the method of choice (spectrophotometric or fluorescence – see Experiment 4). Additional: microscope with oil immersion lens.

Cultures

Ten to fourteen days prior to the exercise, start a nutrient-sufficient culture of the algae of choice. This is done by inoculating 2 L of medium with ca. 200 mL of previously grown log phase cells (dense culture). The culture is best grown in two 2.8L fernbach flasks (1.1 L culture each). The cultures should be either stirred (magnetic stir bar) and/or aerated (filtered air) and grown at ca. 18–20° under ca. 100–150 μE m^{-2} sec^{-1} irradiance in either continuous light or 12:12 light:dark cycle. After 10–14 days (culture should be late-logarithmic growth or early-stationary phase), combine the two volumes. Provide 1 L of this combined culture to the students for the first part of the exercise. Take 1 L of the remaining culture and harvest in a preparative centrifuge (10 min at 3000 × g). Wash the cells twice with nitrate-free medium at the centrifuge by resuspending and centrifuging. Resuspend the cells in 1 L of nitrate-free medium, and set up this culture as previously described. Maintain this culture for approximately four weeks, or until the cells take on a distinctive orange hue. This nutrient-deficient culture serves as the material for this part of the experiment.

Sterile technique is important during the cell wash and transfer cycle, particularly if protease-peptone is added to the medium. Do not aerate cultures if protease-peptone is in the medium.

Algae

A wide range of species are available (see original literature). The sources of the suggested ones are as follows:

UTEX = University of Texas Culture Collection.

CCAP = Cambridge Collection of Algae-Protozoa (Now at the Freshwater Biological Assoc.)

Göttingen = Sammlung von Algenkulturen, Universität Göttingen.

Scenedesmus obliquus (= *S. naegelii*):
    UTEX 74; CCAP 276/2; Göttingen 276-2

Ankistrodesmus braunii
    UTEX 244; CCAP 207/7a

Haematococcus lacustris (= *H. pluvialis*):
    CCAP 34/1c; Göttingen 34-1c

Neochloris wimmeri:
  UTEX 113; CCAP 213/4; Göttingen 213-4
Spongiochloris typica:
  UTEX 1238

Medium

All these cultures can be grown on a basic freshwater medium:

$NaNO_3$	200 mg
KH_2PO_4	50 mg
$MgSO_4 \cdot 7H_2O$	50 mg
$CaCl_2 \cdot 2H_2O$	50 mg
Tris Base	100 mg
Fe solution	2 mL*
Trace elements solution	1 mL**
H_2O	1000 mL

1 g Protease-peptone will improve growth, but can cause problems with sterility.
For nitrogen-deficient medium omit the $NaNO_3$.

*Fe solution			**Trace element solution	
$FeSO_4.7H_2O$	4 g		H_3BO_3	2.9 g
Na_2EDTA	5 g		$MnCl_2 \cdot 4H_2O$	1.8 g
H_2O	1000 mL		$ZnSO_4 \cdot 7H_2O$	2.0 g
			$CuSO_4 \cdot 5H_2O$	80 mg
			$CoCl_2 \cdot 6H_2O$	40 mg
			Ammonium paramolybdate	40 mg
			H_2O	1000 mL

11

Electrophoretic separation and spectral characterization of algal phycobiliproteins

Bruno P. Kremer

Universität zu Köln, Institut für Naturwissenschaften und ihre Didaktik, Abteilung für Biologie, 5000 Köln 41, Federal Republic of Germany

FOR THE STUDENT

Introduction

The dominant accessory pigments in Cyanobacteria, Cryptophyta and Rhodophyta are blue and red compounds. Upon isolation they show a strong fluorescence even in the daylight. They are collectively termed phycobiliproteins. Among them, the red phycoerythrins (PE) occur mainly in red algae (R), while the blue phycocyanins (PC) are preponderantly found in Cyanobacteria. Different phycocyanins and phycoerythrins were originally thought to be characteristic of red algae and bluegreen bacteria. Accordingly, they were named R-PE, C-PE, R-PC, and C-PC.

It has been found, however, that C-PC also occurs in many red algae, while R-PE-like phycobiliproteins (PBP) were consistently reported from a few species of cyanobacteria. A further type of PBP besides PE and PC is allophycocyanin (APC) and this occurs in both cyanobacteria and red algae. On the other hand, phycoerythrocyanin (PEC) is a PBP which has so far been detected only in bluegreen bacteria and there mainly in the heterocyst–forming genera. Originally, the classification of PBP was primarily based upon spectral properties (absorption as well as fluorescence). Modern analysis added further criteria such as subunit composition,

111

amino acid sequences, and immunological details (Ragan 1981; Glazer 1985). All these pigments can easily be separated by electrophoresis.

Phycobiliproteins comprise chromophores and protein (polypeptide) components (Fig. 11.1). The chromophores (= phycobilins) are open-chain tetrapyrrole systems (Fig. 11.2) that are biosynthetically derived from 5-aminolevulinic acid and belong to the same chemical family as chlorophyll and heme. The most frequently occurring phycobilins are phycocyanobilin (PCB) and phycoerythrobilin (PEB). Two further phyco-bilins, phycourobilin (PUB) and a phycobiliviolin–like chromophore (PXB) have been recognised, but have not yet been analysed in greater detail.

One or more phycobilins are covalently bound to polypeptide chains (α-, β-, γ-chains) of different molecular sizes. In most cases, the binding amino acid is cysteine, so that a thioether bridge is formed between cysteine and a side chain of ring A. Such polypeptide–phycobilin associations (Fig. 11.1) form subunits that aggregate into di-, tri- or hexamers (Table 11.1).

In contrast to other chloroplast pigments, phycobiliproteins are not integrated into the thylakoid membranes, but form phycobilisomes on the surfaces of the thylakoids. Phycobilisomes are highly aggregated, supra-molecular complexes that can be observed with the electron microscope. Additional information on phycobiliproteins and phycobilisomes is provided in recent reviews (Gantt 1981; Glazer 1983, 1985; Wehrmeyer 1983).

Phycobiliproteins dominate the pigmentation of cyanobacteria, cryptomonads and red algae. Their absorption maxima range between 500 and 650 nm. They thus fill the 'green window' between the absorption peaks of chlorophyll *a*. Consequently, organisms equipped with PBP can effi-

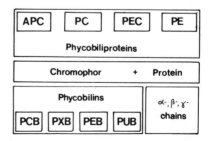

Fig. 11.1. Schematic survey of types, components and composition of phycobiliproteins. Abbreviations are explained in the text.

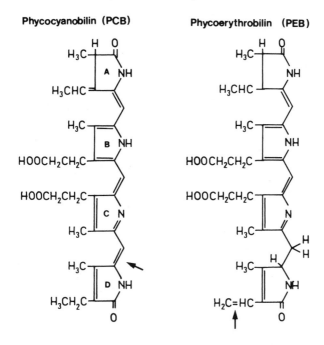

Fig. 11.2. Chemical structure of phycocyanobilin (PCB) and phyco-erythrobilin (PEB). The compounds are isomers of each other. Arrows indicate structural differences.

ciently use those quanta of photosynthetically active radiation that are only inefficiently absorbed by most of the green plants.

In this experiment you will extract and separate the phycobiliproteins from cyanobacteria, cryptomonads and/or red algae, using agarose gel electrophoresis. In addition, the absorption spectra of the separated PBPs are determined.

Procedures (Fig. 11.3)

Extraction. Homogenize about 20 g fresh weight or 5 g freeze-dried algae by grinding in a mortar (preferably in liquid N_2) and then extract in about 100 mL extraction buffer. Homogenization can be speeded up with ultrasound (in cold) or a French pressure cell (in cold, only with unicellular organisms). Be careful about high voltage if you are using liquid N_2! Extract the homogenate by stirring continuously overnight (12–17 h) at 5°C (refrigerator).

Table 11.1. *Survey and composition of phycobiliproteins (PBP). For abbreviations see text*

PBP	Absorption maxima (nm)	Most frequent aggregation of subunits at pH 7	Distribution of Chromophores on single subunits			Occurrence		Example
			α	β	γ	Cyano-bacteria	Red algae	
APC	650	$(\alpha\beta)_3$	1 PCB	1 PCB	—	+		*Anacystis nidulans* / *Porphyridium purpureum*
C-PC	620	$(\alpha\beta)_6$	1 PCB	2 PCB	—	+	+	*Fremyella diplosiphon* / *Porphyridium aerguineum*
R-PC	620, 555	$(\alpha\beta)_3$	1 PCB	1 PCB 1 PEB	—		+	*Porphyridium purpureum* / *Palmaria palmata*
PEC	575, 590(s)[1]	$(\alpha\beta)_3$	1 PXB	2 PCB	—	+		*Anabaena variabilis* / *Mastigocladus laminosus*
C-PE	565	$(\alpha\beta)_6$	2 PEB	3–4 PEB	—	+		*Fremyella diplosiphon* / *Nostoc muscorum*
B-PE	545, 565, 495(s)	$(\alpha\beta)_6\gamma$	2 PEB	4 PEB	2 PEB 2 PUB		+	*Porphyridium purpureum* / *Rhodella violacea*
R-PE	565, 540, 495	$(\alpha\beta)_6\gamma$	2 PEB (?)[2]	4 PEB (?)	2 PEB 2 PUB (?)		+	*Ceramium rubrum* / *Delesseria sanguinea*

[1] (s) = Shoulder
[2] Similar to B-PE. Exact number of chromophores not yet determined

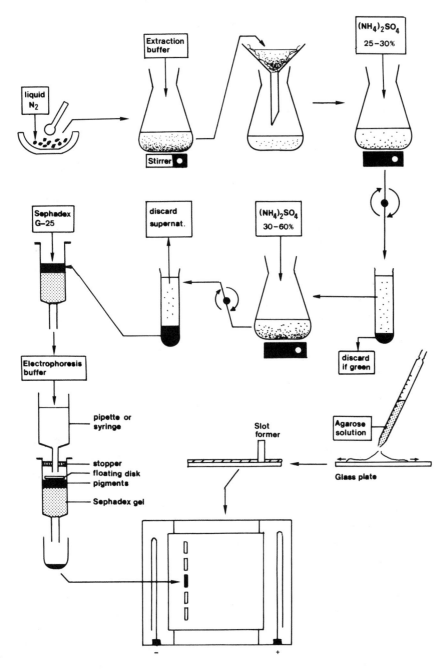

Fig. 11.3. Flow chart showing extraction, purification and electrophoretic separation of phycobiliproteins.

Purification. Filter the resulting suspension through a multiple layer of cheese cloth and a paper filter. Centrifugation (5 min at 2000 × g) removes any particulate material that passed the filters.

The PBP-containing supernatant (= extract) must now be further purified. This is accomplished through a stepwise fractionation procedure by adding increasing amounts of solid ammonium sulfate to the solution. This will finally salt out the phycobiliproteins. The precipitation is performed at 5°C with continuous stirring.

In a first stage, add 16.5 g $(NH_4)_2SO_4$ in portions of about 2 g min^{-1} to your 100 mL pigment extract to give 30% saturation. Then stir in the cold for an additional 30 min. After this extraction step, a greenish precipitate (chlorophyll-protein complexes) is found. Discard by centrifugation (5–10 min at 2000 × g). Then add further 20 g $(NH_4)_2SO_4$ (as per previous procedure) to bring to 60% saturation. In the saturation range between 30%–60% $(NH_4)_2SO_4$, blue, reddish or even violet material will precipitate, representing the precipitated phycobiliproteins desired for the next steps. Collect by centrifugation (5–10 min at 2000 × g) and discard supernatant.

Resuspend the precipitated pigments in a small volume (1–3 mL) of extraction buffer. Insoluble material is removed by centrifugation. Now the pigments must be deionized or rebuffered. This is achieved in minicolumns made of syringes and filled with Sephadex G-25 gel (prepared by the instructor). Sephadex G-25 is commonly used for desalting or for buffer changes in proteinaceous solutions.

Place up to 1 mL of the concentrated phycobiliprotein solution on the gel surface by means of a bent Pasteur pipette. The pigments are then chromatographed and eluted with electrophoresis buffer. A floating polyethylene disc prevents disrupting of the gel surface by dropping buffer.

Preparation of agarose plates. In a 100 mL Erlenmeyer flask, prepare a 1% agarose solution by suspending the agarose at room temperature in electrophoresis buffer. Melt this suspension in a boiling-water bath, stirring continuously. Some 25 mL of the warm agarose solution is run out onto horizontal glass plates (9 × 11 cm or 10 × 10 cm), and the slot-former is immersed into the still warm agarose. It provides the slots (1.5 × 20 mm or 2.0 × 15 mm) for the application of the pigment samples. Remove the slotformer about 30–45 min later after cooling and setting of the agarose gel. Store the agarose plates in a humidity chamber (plastic food storage box) at 5°C for 1–2 h before use.

Electrophoresis. The electrophoretic separation is carried out using the submerged-gel technique with submersed agarose gel. The complete electrophoresis unit must be placed horizontally. Then fill each buffer chamber with 750 mL electrophoresis buffer. Cover also the agarose plate (application slots toward the cathode) completely with electrophoresis buffer (Fig. 11.4).

Add solid sucrose (to about 10% w/v) to the purified phycobiliprotein solutions and dissolve; this increases the density of the solutions and prevents the pigment samples from floating out of the slots. Carefully pipette about 50–100 μL of phycobiliprotein solution into the buffer-covered slot by means of a 0.5 mL or 1 mL micropipette.

The separation is performed with 50 mA D.C. in cold buffer (10°C) and

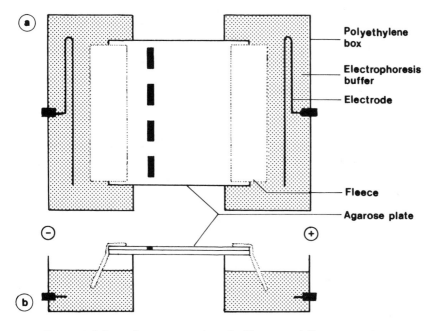

Fig. 11.4. Schematic representation of a "home-made" apparatus for electrophoretic separations. The electrode chambers (buffer chambers) are slightly modified polyethylene boxes (e.g. tupperware boxes). The connection between the chambers and the agarose gel is provided by appropriate pieces of filter paper or cellulosic fleece. a = surface view, b = side view.

will require about 2–3 h. Caution is recommended with the high voltage! You can transfer the whole unit to a refrigerator after the pigment complexes have entered the agarose gel (about 15 min after start). This will ensure a suitable temperature range during the course of the separation. Alternatively, a cooling plate, often available in commercial electrophoresis equipment, can be applied.

During electrophoresis, the initially homogenous pigment samples separate into several bands depending on the source. The different classes of phycobiliproteins (PE, PC, APC, PEC) are readily distinguishable by their characteristic coloration and their different electrophoretic mobility. Generally, the brilliantly red colored PE bands prove to be the fastest moving on the agarose plates, very clearly separated from the intensive blue PC or the light blue APC (often only detectable upon staining, see below). PEC exhibits the lowest electrophoretic mobility.

In the same way that R_f values can be used to characterize the mobility of particular compounds on thin-layer or paper chromatograms, the position of a specific band can be expressed by its electrophoretic mobility (R_e). This value is obtained by the following equation:

$$R_e = \frac{\text{distance compound traveled (mm)}}{\text{distance fastest moving compound traveled (mm)}}$$

Spectrophotometry. The intensively colored bands can be re-eluted from the agarose gels for further analysis. Cut off the appropriate zone of the gel with a razor blade and suspend it in extraction buffer. Use the resulting liquid to characterize the spectral properties of the respective compound by measuring its absorption (extinction) for wavelengths between 400 and 700 nm. For this purpose, the use of a recording spectrophotometer (if available) is preferable.

Preservation and staining. You can preserve the brilliantly colored pattern of diverse phycobiliproteins obtained by electrophoresis. Simply dry the agarose to thin films on their supporting glass plates at 60°C for 2 h in an oven. The dried agarose is then easily detached from the glass. Remember that the isolated phycobiliproteins are still senstive to bright sunlight. Fully exposed, they will fade completely within a few days.

Before drying, the agarose layers can also be stained with 0.05% Serva blue W (aqueous): Cover the gel with the stain for 30 min, wash with several changes of distilled water (5–10 min) and finally dry at 60°C. Such a staining very much intensifies the bands of minor phycobiliproteins and also allows for the detection of any uncolored proteins besides the phycobiliproteins, if they should be present.

Questions

1. What does the electrophoretic separation of a complex phycobiliprotein sample into several distinguishable bands depend on?
2. Which component of the phycobiliproteins could determine the overall electrical charge of the molecules and their resulting mobility in an electrical field?
3. What differences are there between the species analyzed?
4. How many bands of phycobiliproteins did you detect?
5. Are there species possessing mainly or exclusively PE or PC? Where did you trace PEC?
6. Does the relative intensity of the phycobiliprotein bands reflect the approximate proportions of PE, PC and APC? Could you determine the relative amounts present?
7. Does the pattern of phycobiliprotein bands correspond with any taxonomic concept? Can you identify similarities or differences between cyanobacteria, cryptomonads and red algae?
8. Would you expect the pattern of bands to be constant within different samples from the same species, within the same genus, family, order?
9. Compare the extraction procedure performed with that recommended for chlorophylls and carotenoids (Experiment 9). Where are the differences? Which special features of the pigments do they reflect?
10. Why can phycobiliproteins not be separated by thin-layer chromatography and chlorophylls or carotenoids (usually) not by electrophoresis?
11. Compare the chemical structure of the chromophore of phycobiliproteins and chlorophylls/carotenoids. In which groups do you suspect common biosynthetic precursors?

REFERENCES

Gantt, E. (1981). Phycobilisomes. *Ann. Rev. Plant Physiol.* 32, 327–347.

Glazer, A. N. (1983). Comparative biochemistry of photosynthetic light-harvesting systems. *Ann. Rev. Biochem.* 52, 125–137.

Glazer, A. N. (1985). Light harvesting by phycobilisomes. *Ann. Rev. Biophys.* 14, 47–77.

Lobban, C. S., Harrison, P. J. & Duncan, M. J. (1985). *The Physiological Ecology of Seaweeds.* Cambridge University Press, Cambridge, 242 pp.

Ragan, M. A. (1981). Chemical constitutents of seaweeds. In, C. S. Lobban & M. J. Wynne (eds.) *The Biology of Seaweeds.* Blackwell Scientific Publications, Oxford, pp. 589–626.

Wehrmeyer, W. (1983). Organization and composition of cyanobacterial and rhodophycean phycobilisomes. In, G. C. Papageorgiou & L. Packer (eds.) *Photosynthetic Prokaryotes: Cell Differentiation and Function.* Elsevier Science Publishing Co., Amsterdam, pp. 1–22.

NOTES FOR INSTRUCTORS

Several methods are available for the separation of phycobiliproteins eluted from a Sephadex column. Besides gel filtration and ion exchange chromatography, polyacrylamide gel electrophoresis or isoelectric focusing may be successfully used. In our experiment, the least complicated method, flatbed electrophoresis in agarose gels, is selected.

Organisms

Biological material includes any cyanobacterial, cryptomonadal and red algal samples available in quantities of 20 g (fresh weight) or 5 g (dry weight; freeze-dried and stored in darkness).

Equipment

1. magnetic stirrer
2. refrigerator
3. oven (optional)
4. cooling laboratory centrifuge
5. power unit providing 4–6 mA and 120–160 V direct current (field strength 7.5–10 V cm^{-1})
6. electrophoresis: commercially-made flatbed system or electrophoresis chambers assembled from household boxes (tupperware boxes about 20 × 10 × 5 cm) (Fig. 11.4). For safety, the electrophoresis chambers should be placed on a shallow (ca. 2 cm) tray with rubber feet as insulation.
7. slotformer: cut out several small strips of acrylic glass (10 × 10 mm in section) and glue them together according to Fig. 11.5. Strips of the following lengths are needed: 150 mm (1), 60 mm (2), 100 mm (2). Additionally, eight smaller pieces (15 × 10 × 2 mm) are required; four of them are glued on the lower surface of the 100 mm-strip. These pieces are submerged into the still warm agarose gel and thus form the slots for the pigment extracts.
8. spectrophotometer (recording type, optional)

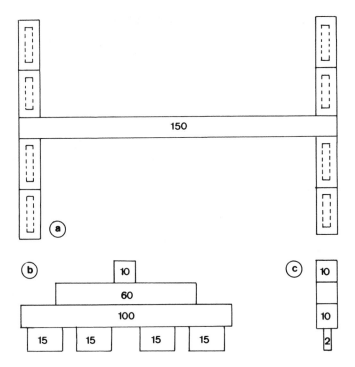

Fig. 11.5. Schematic representation of a slotformer constructed from small strips and pieces of acrylic glass. All dimensions in mm; a = surface view, b = front view, c = side view.

Glassware (per group of 2–3 students)

1. Erlenmeyer flasks (per algal sample 1 × 250 mL, 3 × 100 mL)
2. centrifuge tubes (per algal sample 1 × 100 mL or 2 × 50 mL or 4 × 25 mL according to centrifuge type)
3. pipettes 1, 2, 5, 10, 25 mL
4. tuberculin syringes or 0.1 mL pipettes
5. some glass wool, filter paper, cheese cloth
6. funnel
7. mortar and pestle
8. glass plates 9 × 11 cm or 10 × 10 cm

Chemicals (per group of students)

1. Extraction buffer: 0.1 M phosphate buffer, pH 6.9, including 1 mM EDTA (free acid) and 1 mM dithiothreitol. Preparation: Dissolve 9.08 g KH_2PO_4 in 1000 mL (= A) and 11.88 g $Na_2HPO_4.2H_2O$ in 1000 mL

($= B$); mix 1000 mL A and 710 mL B and add 500 mg EDTA and 262 mg dithiothreitol. Store in refrigerator.

2. Electrophoresis buffer: 0.125 M Tris/H_3BO_3, pH 8.9. Preparation: Dissolve 15.1 g Tris(hydroxymethyl)aminomethane and 7.7 g H_3BO_3 in 1000 mL. Cool to 10°C.
3. $(NH_4)_2SO_4$ (0.5 kg)
4. Sephadex G-25 (medium grade)
5. Serva blue W (Serva No. 35053) (1 g)
6. Agarose without EEO (electroendosmose) (e.g. Serva No. 11401) (5 g)
7. liquid N_2 (optional)

Syringe columns for deionization: In order to save class time, a sufficient number (one per pigment sample) of packed syringes should be prepared in advance. Sephadex G-25 powder is swollen and washed with 5–7 volumes of distilled water for 1–2 h. The piston of a disposable 10 mL syringe is removed and some glass wool layered on its outlet to retain the gel. Fill this device with Sephadex G-25 gel. Allow 50 mL electrophoresis buffer to pass through this column for equilibration. Store (for max. 2–3 d) in a refrigerator until use.

Scheduling

The scheduling of this experiment could be as follows:

— extraction	overnight (12–17 h)	
— purificaiton	1–2 h	Simultaneously by
— preparation of		different students
agarose gel plates	1–2 h	of the same group
— spectrophotometry	1 h	
— preservation	1 h	
— discussion	1 h	

The main experimental steps (extraction, purification, separation, etc.) can be performed on several subsequent days, since all materials can be stored at 5°C for up to one week.

   Each group of students should process at least two or even four different algal samples to provide a sufficient data basis for a comparative discussion of the results.

12

The effects of spectral composition and irradiance level on pigment levels in seaweeds

L. V. Evans

Department of Plant Sciences, University of Leeds, Leeds LS2 9JT, United Kingdom

FOR THE STUDENT

Introduction

Variations occur in the photosynthetic pigments (chlorophylls, carotenoids and phycobiliproteins) in the different classes of algae. The Chlorophyceae have chlorophylls a and b and carotenoid accessory pigments (carotenes and xanthophylls) like those of higher plants. In the Phaeophyceae and Rhodophyceae, chlorophyll b is absent. In the former it is replaced as an accessory pigment by chlorophylls c_1 and c_2 and, although β-carotene is present in low concentrations, the major carotenoid is the photosynthetically-active xanthophyll fucoxanthin. In the Rhodophyceae, although chlorophyll d is said to be present (this is not proved, and may be an artefact) and xanthophylls such as lutein are present in some species and carotenes occur in all species, the major accessory pigments are the phycobiliproteins phycoerythrin (PE) and phycocyanin (PC). The principal absorption bands of these pigments are shown in Table 12.1.

Because both the intensity and the spectral composition of light change with increasing depth of water, the characteristics of the light received by a seaweed depend to a large extent on the depth at which it is growing (see e.g. Dring 1982; Lobban et al. 1985; papers by Ramus and Ramus and colleagues). Variation in pigment composition to make optimal use of the

123

Table 12.1 *Principal absorption peaks (in nm) of the main pigments of brown and red algae in organic solvents*

Chlorophylls	
chl *a*	420, 662
chl *b*	455, 644
chl *c*	444, 626
chl *d*	450, 690
Carotenoids	
α-carotene	420, 440, 470 (in hexane)
β-carotene	425, 450, 480 (in hexane)
lutein	425, 445, 475 (in ethanol)
fucoxanthin	425, 450, 475 (in hexane)
Phycobiliproteins	
phycoerythrins	495, 545, 564 (main)
phycocyanins	618
allophycocyanin	650

prevailing light is therefore very important. However, the question of whether it is the difference in light intensity (irradiance) or the difference in light quality (spectral distribution) that is primarily responsible for the pigment changes is one that has been much debated. Engelmann (1883), in his theory of chromatic adaptation, proposed that green, brown and red seaweeds predominate at different depths because their pigment composition is complementary to the quality (spectral composition) of the prevailing light. As light becomes greener with depth, the presence of fucoxanthin (brown algae) and phycoerythrin (red algae) should become increasingly advantageous. Oltmanns (1892), on the other hand, was of the view that irradiance level was more important than spectral composition, and Dring (1981) also concluded that the pigment composition changes associated with increasing depth in seaweeds are largely adaptations to low irradiance rather than to the different spectral composition of the light. Although the former may be the chief factor, however, it is likely that some degree of adaptation to light quality may be superimposed upon this (for further discussion see Ramus references). An example of chromatic adaptation is studied in Experiment 13.

In this experiment, data will be obtained from which you can make your own evaluation of the relative importance of irradiance level and spectral distribution upon the pigment composition of brown and red seaweeds. (For chlorophyll analysis of green algae, see Experiment 4A.)

Procedure

Use small plants, or apices, thallus pieces or discs taken from brown and/ or red seaweeds that have been grown under white, blue, red and green light fields at two irradiance levels (e.g., 30 and 300 μE m^{-2} s^{-1}) at approximately 10°C for a period of 2–4 weeks.

CAUTION! WEAR GLOVES WHEN HANDLING DMSO. It is very readily absorbed through skin and takes contaminants with it.

For a thalloid brown seaweed (after Seely et al., 1972).
Keep everything in dim light as far as possible. See Precautions in Experiment 9 (but do not keep DMSO on ice.)

1. Rinse material (e.g., one apex or one 10–15 mm disc) in tap water to remove salt water, blot dry, weight out 0.05–0.5 g, chop into small pieces (1–2 mm square), transfer to small snap-top tubes and extract in 0.4 mL dimethyl sulfoxide (DMSO) for 15 min, swirling occasionally. Keep everything in darkness if possible.

2. Pipette off the bright yellow extract and wash the pieces twice (5 min each) in 0.3 mL DMSO, swirling occasionally. Add this DMSO to that from Step 1, check total volume if possible and carry out spectrophotometry using formulae as in Step 7. (a), using microcuvettes.

3. Extract five times (5–10 min each, or until acetone extract is clear) in 0.6 mL 90% acetone. Pieces should now be white. (Grinding of tough material may be necessary.) The acetone extracts contain fucoxanthin and chlorophyll c together with β-carotene and chlorophyll a. The former are now separated from the latter by partitioning and extraction, as described in Steps 4–6 and summarised in Fig. 12.1.

4. Combine the acetone extracts (total 3 mL), then add to the combined extracts 1 mL hexane and 1 mL distilled water. Swirl gently, but avoid shaking. For small volumes, this may be done in snap-cap tubes; for larger volumes use, e.g., a 50 mL separating funnel.

5. Using a Pasteur pipette, or by running off the lower acetone extract from a separating funnel, separate the phases. Then wash the upper hexane phase 2–3 times (or until washings are colorless) with an equal volume of 80% methanol. Make a note of the total volume of methanol used. Keep the washed hexane phase for Step 6. Combine the methanol washings with the aqueous phase of the acetone extract from Step 4 and carry out spectrophotometry (with glass cuvettes), using formulae as in Step 7 (b) below. Read three samples of your extract and take an average. (A small hexane-rich phase may appear on the surface; if so, separate and combine it with the main hexane phase.)

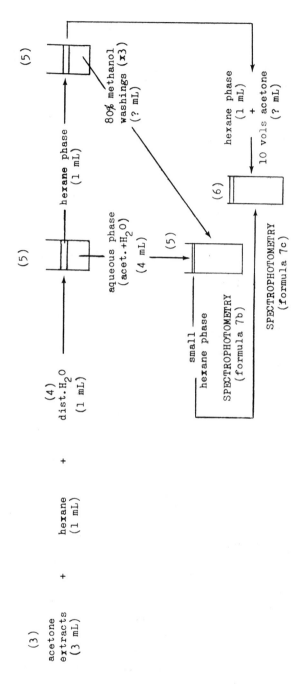

Fig. 12.1. Summary of pigment extraction/separation protocol for brown seaweeds.

6. Dilute the (1 mL) hexane phase with 10 volumes of acetone. Carry out spectrophotometry with glass cuvettes, using formulae as in Step 7 (c). Read three samples of your extract and take an average.

7. Determine the pigment concentrations in the extracts, using the following formulae. Concentrations are expressed in g L^{-1}. Chl a = chlorophyll a, fx = fucoxanthin, car = β-carotene. (Note that solutions for spectrophotometry should be clear; if cloudiness occurs in the extracts at Steps 5 and 6, this can be eliminated by passage through NaCl or anhydrous Na_2SO_4 or, for the acetone:methanol:water extract by adding more methanol.)

(a) DMSO Extract

[chl a] = $A_{665}/72.8$

[chl c] = $(A_{631} + A_{582} - 0.297\ A_{665})/61.8$

[fx]　　= $(A_{480} - 0.722\ (A_{631} + A_{582} - 0.297\ A_{665}) - 0.049\ A_{665})/130$

(b) Acetone:Methanol:Water Extracts

[chl a] = $A_{664}/73.6$

[chl c] = $(A_{631} + A_{581} - 0.300\ A_{664})/62.2$

[fx]　　= $(A_{470} - 1.239\ (A_{631} + A_{581} - 0.300\ A_{664}) - 0.0275\ A_{664})/141$

(c) Acetone:Hexane Extracts

[chl a]= $A_{661}/83.3$　　(or $A_{615}/15.4$)

[car]　= $(A_{480} - 0.033\ A_{661})/193$

8. Next, determine the total amount of each pigment extracted. Bear in mind that you cannot add concentrations determined from (a), (b) and (c) directly because different volumes are involved. First convert these concentrations to amounts of pigments actually present in each extract, then add to get a total of each pigment extracted from the tissue.

9. Calculate all pigment concentrations in mg g^{-1} f.w. of material. In addition, for flat-blade algae, it is also sometimes helpful to express in mg $area^{-1}$.

10. Logically present all data for the material you yourself have investigated, showing clearly how the totals have been calculated for each pigment.

11. Tabulate the results for all the material investigated by the class, and discuss.

For a thalloid red seaweed:

(a) Chlorophyll and carotenoids (after Seely et al., 1972):

1. Proceed as in Steps 1 and 2 in the previous section (note that in Step 2 the red algal extract will not be bright yellow), and carry out spectrophotometry for chlorophyll a only, as in Step 7 (a) earlier.

2. Extract five times (5–10 min each, or until acetone extract is clear) in 0.6 mL 90% acetone. Using glass cuvettes, carry out spectrophotometry for chlorophyll *a* and carotenoids as in Step 7 (c) in the previous Section. Read three samples of your extract if possible and take an average.

(*b*) *Phycobiliproteins* (after Beer & Eshel, 1985):

1. Weigh out another 0.05–0.5 g of thallus (some loss of water-soluble pigment may have occurred during extraction of the material in Steps 1-2 above), transfer to a mortar and grind thoroughly in 5 mL 0.1 *M* phosphate buffer, pH 6.8, with a little acid-washed sand. Centrifuge (add another 5 mL phosphate buffer if necessary, to increase liquidity) for 10 min at 1,000 × g (usually maximum speed). Transfer supernatant to 25 mL volumetric flask. Repeat grinding/centrifugation until no further pigments can be extracted, adding supernatants to the contents of the volumetric flask. Make up to 25 mL accurately. (Centrifuge for 15 min at top speed on bench centrifuge if cloudy. However, to ensure routinely clear solutions for spectrophotometry, refrigerated centrifugation for 10–15 min at ca. 20,000 × g is recommended.)
2. Determine phycobiliprotein concentrations using the following formulae. Concentrations are expressed in mg mL^{-1}; R-PE = phycoerythrin, R-PC = phycocyanin.

$$[\text{R-PE}] = [(A_{564} - A_{592}) - (A_{455} - A_{592})\ 0.20]\ 0.12$$
$$[\text{R-PC}] = [(A_{618} - A_{645}) - (A_{592} - A_{645})\ 0.51]\ 0.15$$

3. Proceed as in Steps 8–11 earlier.

Questions

1. In extraction of pigments, what does DMSO do?
2. Why are the estimations of pigments extracted here not absolute?

REFERENCES

Beer, S. & A. Eshel (1985). Determining phycoerythrin and phycocyanin concentrations in aqueous crude extracts of red algae. *Aust. J. Mar. Freshw. Res.* 36, 785–792.

Dring, M. J. (1981). Chromatic adaptation of photosynthesis in benthic marine algae: an examination of its ecological significance using a theoretical model. *Limnol. Oceanogr.* 26, 271–284.

Dring, M. J. (1982). *The Biology of Marine Plants.* Edward Arnold, London, 199 pp.

Dring, M. J. (1986). Pigment composition and photosynthetic action spectra of sporophytes of *Laminaria* (Phaeophyta) grown under different light qualities and irradiances. *Br. phycol. J.* 21, 199–207.

Engelmann, T. W. (1883). Farbe und Assimilation. *Bot. Zeit.* 41, 1–29.

Gallagher, J. C., A. M. Wood & R. S. Alberte (1984). Ecotypic differentiation in the marine diatom *Skeletonema costatum:* influence of light intensity on the photosynthetic apparatus. *Mar. Biol.* 82, 121–134.

Kursar, T. A. & R. S. Alberte (1983). Photosynthetic unit organization in a red alga. *Plant Physiol.* 72, 409–414.

Lobban, C. S., P. J. Harrison & M. J. Duncan (1985). *The Physiological Ecology of Seaweeds.* Cambridge University Press, Cambridge, 242 pp.

Oltmanns, F. (1892). Uber die Kultur und Lebensbedingungen der Meeresalgen. *J. Wiss. Bot.* 23, 349–440.

Ramus, J. (1983). A physiological test of the theory of complementary adaptation. II. Brown, green and red seaweeds. *J. Phycol.* 19, 173–178.

Ramus, J., S. I. Beale & D. Mauzerall (1976). Correlative changes in pigment content with photosynthetic capacity of seaweeds as a function of water depth. *Mar. Biol.* 37, 231–238.

Ramus, J., F. Lemons & C. Zimmerman (1977). Adaptation of light-harvesting pigments to downwelling light and the consequent photosynthetic performance of the eulittoral rockweeds *Ascophyllum nodosum* and *Fucus vesiculosus. Mar. Biol.* 42, 293–303.

Ramus, J., S. I. Beale., D. Mauzerall & K. L. Howard (1976). Changes in photosynthetic pigment concentration in seaweeds as a function of water depth. *Mar. Biol.* 37, 223–229.

Seely, G. R., M. J. Duncan & W. E. Vidaver (1972). Preparative and analytical extraction of pigments from brown algae with dimethyl sulfoxide. *Mar. Biol.* 12, 184–188.

NOTES FOR INSTRUCTORS

Materials

For pretreatment of material (2–4 weeks before date of practical):

1. Suitable seaweed material (see comments that follow)
2. Suitable culture vessels (see comments that follow) and an aeration pump
3. A good light source, e.g., a high-intensity mercury lamp (such as used in commercial greenhouses), or white fluorescent lamps, if sufficient are available to give the required upper irradiance level when filters are used.
4. An angle-iron stand with two glass shelves, so that culture vessels may be positioned at two distances from the light source.
5. A constant temperature room able to run at $10 \pm 2°C$, even when the lamp(s) is (are) on.
6. Green, blue and red cellophane, gelatin or perspex filters of known spectral characteristics, as used for TV and stage lighting (see e.g., Dring, 1986).
7. Material to reduce irradiance level without affecting spectral com-

position, e.g., muslin, window screen or neutral (white coloured) TV lighting material.

8. Spectrophotometer (recording, if possible) to check absorbance of all filters.
9. Quanta meter.
10. Erdschreiber or other culture medium suitable for marine macroalgae

For experiment:

1. laboratory balance(s)
2. spectrophotometer(s)
3. bench centrifuge
4. snap-cap tubes (10 per student)
5. 50 mL separating funnels (1 per student)
6. microcuvettes (1 mL volume)
7. glass cuvettes
8. graduated or automatic pipettes (0.3 mL, 0.4 mL, 0.6 mL and 1.0 mL volumes to be dispensed)
9. Pasteur pipettes
10. graduated measuring cylinder (10 mL)
11. volumetric flask (25 mL)
12. mortar and pestle (1 per student if extracting red seaweeds) – porcelain if using liquid N_2.
13. acid washed sand (for red seaweed)
14. liquid N_2 (if algal tissue is very tough)
15. DMSO (10 mL per student)
16. 90% acetone (10 mL per student)
17. distilled water
18. technical grade hexane (10 mL per student), or petroleum ether 60–90° fraction
19. 80% methanol (25 mL per student)
20. 0.1M phosphate buffer, pH 6.8 (50 mL per student), if using red algae

Comments

Pretreatment of material. Suitable brown seaweeds include *Fucus* species (apices), *Laminaria, Dictyota* and *Zonaria* species (10–15 mm discs). Where such seaweeds cannot be collected or purchased from a supplier, substitute an alga with a similar basic pigment composition that can be cultured, e.g., *Skeletonema* (see e.g., Gallagher et al., 1984). Suitable red algae include *Palmaria palmata* (10–15 mm discs), *Chondrus crispus, Gracilaria* spp, *Pterocladia capillacea, Jania* and *Laurencia* species (thallus pieces/branches). Red algae that could be tested for use after growth in

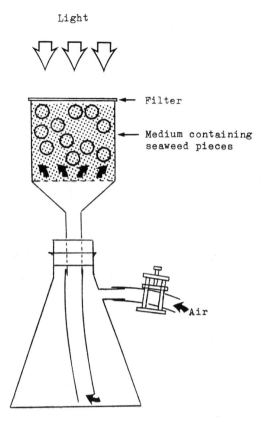

Fig. 12.2. Apparatus used for growing seaweed pieces under different lighting conditions. Air pumped through the system (arrows) keeps the seaweed pieces in constant motion.

culture include *Rhodella, Porphyridium, Bangia* and *Rhodochorton* species.

Material can be maintained during pretreatment in suitable vessels containing e.g., Erdschreiber medium, aerating if possible (see e.g., Dring 1986). In the author's laboratory, 11 cm diam. straight-sided 500 mL sintered funnels are used. These are set up on 500 mL conical side–arm flasks, and pieces of polythene tubing are attached to their stems of sufficient length to reach the bottom of the flask (see Fig. 12.2). Air is driven from a pump through the side-arms (one pump will support many flasks) and passed through the sintered base of the funnel upwards through the contained medium, keeping the seaweed pieces in motion. Gate clips are used to control air flow into each flask. The incubation vessels used should have to control air flow into each flask. The incubation vessels used should have glass lids over which the filters are placed. The sides of the vessels should be enclosed in tinfoil.

Filters (green, blue and red) of different materials, as used for stage or TV lighting are widely available (e.g., Edscorp filters, with known transmission characteristics). It is advisable to check (using a spectrophotometer) that the filters are absorbing only the intended wavelengths. The highest photon irradiance that can be used is dependent on the light source (a high-intensity mercury lamp, as used in commercial greenhouses, is ideal) and by the filter that has the lowest overall transmission, as assessed by a quanta meter. The irradiance level has to be the same in all vessels, so that the levels of the others have to be reduced accordingly by careful positioning under the lamp or by using neutral material (i.e., one that lowers the irradiance level without affecting spectral composition). If commercially-supplied neutral filters are used, their absorbance should be checked, as they sometimes affect some spectral regions more than others.

It is advisable to preincubate (start 2–4 weeks before practical) in a constant temperature room capable of maintaining a temperature of $10 \pm 2°C$ when the lamp is on. The lower irradiance level should be 10 times less than the higher level. In the author's laboratory, 35 and 350 μE m^{-2} s^{-1} are used. It is crucial to maintain healthy tissue, completely free of epiphytes of all kinds.

Pigment extraction (1) Gloves should be worn for DMSO and acetone extractions. (2) For spectrophotometric reading of DMSO extractions, 1 mL microcuvettes are ideal; if these are not available, the extract can be diluted with 80% DMSO to make the required larger volume. Larger volumes of DMSO may also be used at each step for the initial extractions, in which case particulate matter may be removed by centrifugation or filtration before finally making up to a known volume for spectrophotometry. If larger volumes of DMSO are used for the initial extraction, it is recommended that the volume of DMSO be equal to four times the weight of tissue. Thallus pieces could be extracted in e.g., an evaporating dish, using larger volumes of solvents, and finally making up to a known volume. Dilution may be necessary before spectrophotometry. Solution concentration will depend on choice of alga. The Instructor may therefore need to try out a few species to establish the best choice, as well as the amount of material to use. In general it is better to err on the side of too much, diluting later by a known amount if necessary. (3) For particularly tough material (e.g., *Fucus*) it may be helpful to grind first in liquid N_2 or acid washed sand, or to put in solvents in snap-cap tubes in an incubator at 30°–40°C. The extraction times may also need to be extended. (4) Allophycocyanin is not usually detectable in crude algal extracts at 650 nm. (Beer & Eshel [1985] found this wavelength represented a trough between PC and H_2O-extractable chlorophyll *a*, which peaked at 675 nm.) The formulae given are for R-PE. Since this is the dominant form in most red algae, the absorption and influence of other types of PE (i.e., B- or C-PE) are not considered here. (See, however, e.g., Kursar & Alberte 1983 for a workable R-PE equation in the presence of C-PC, and an equation for the latter.)

Scheduling. Details of running the experiment depend on the numbers of students. With a class of 16, the author uses two seaweeds (one brown, one red), grown under green, blue, red, and white light at two different irradiance levels. Within a 3 h practical, each student concentrates on one of these, doing as many replicates as possible (up to five pieces of seaweed are available for each), so that simple statistics (average values, standard deviation) may be used. In general, it is better to increase numbers of replicates rather than number of different samples/treatments. All final data are made available to all class members.

13

Chromatic adaptation in Cyanobacteria

David J. Chapman

Dept. of Biology, University of California, Los Angeles, CA
90024, USA

FOR THE STUDENT

Introduction

Phycobiliproteins are a specific group of chromoproteins composed of one or more tetrapyrrole pigments covalently bound to peptide chains (Gantt 1981; Scheer 1982; Glazer 1982). These proteins are found in Cyanobacteria, Rhodophyceae, Glaucophyceae and Cryptophyceae. In all but the latter the phycobiliproteins are located in supramolecular complexes called phycobilisomes (Gantt 1980; Glazer 1984). These pigments are photosynthetically active, and light energy absorbed by them is transferred to the reaction centers of photosystems II and I.

It is the presence of these pigments in large amounts (up to 60% of cell protein) that imparts the characteristic color to the cells. In the Cyanobacteria there are three main classes of these chromoproteins; the blue phycocyanins (C-phycocyanin and allophycocyanin), which are found in all Cyanobacteria; the red phycoerythrin (e.g., C-phycoerythrin) and the purplish phycoerythrocyanin, which are found in some strains. In a series of experiments and observations between 1883 and 1906, Engelmann and later Gaidukov observed that the pigmentation of these algae is influenced by the light quality. This phenomenon, called complementary chromatic adaptation, occurs because the quality of light regulates the synthesis of certain phycobiliproteins (Bogorad 1975; Tandeau de Marsac 1983). Complementary chromatic adaptation is not observed in all Cyanobacteria. In those strains in which it does occur, two types have been observed: (1) where the light regulates phycoerythrin synthesis, and (2) where light regulates both phycoerythrin and phycocyanin synthesis. In both types, green

134

light enhances the synthesis of phycoerythrin, and additionally in type (2) red light lowers the synthesis of phycoerythrin and promotes that of phycocyanin. The net result is a change in the phycobiliprotein content and the resulting color of the cells. Chromatic adaptation is readily observed with two Cyanobacteria, *Tolypothrix tenuis* (Ohki et al. 1985) and *Fremyella diplosiphon* (Bennett and Bogorad 1973; Beguin et al. 1985; Rosinski et al. 1981), both of which contain chlorophyll *a*, allophycocyanin, C-phycocyanin and C-phycoerythrin.

In this experiment you will evaluate the effect of light quality on phycobiliprotein biosynthesis and content in *Fremyella diplosiphon*.

Procedure

You will be provided with a culture (600 mL) of *Fremyella diplosiphon*, which has been grown in cool white light (from fluorescent tubes).

Growth of cells

1. Examine the morphology of the cells in the filaments. What is their morphology and general shape?
2. Divide the culture into five equal aliquots.
3. Transfer three of these aliquots to each of three fernbach flasks containing 1 L of fresh additional medium. These three cultures will be maintained for three weeks under one of the following three light regimes.

   #1 Cool white fluorescent light

   #2 Red light (Red cellophane filters)

   #3 Green light (Green cellophane filters).

Dry weight determination

4. Take the fourth aliquot and centrifuge the cell mass into a weighed centrifuge tube. Centrifuge in a clinical centrifuge at full speed for 10 minutes.
5. Wash the cell pellet twice with distilled water at the centrifuge.
6. Cover the mouth of the tube with aluminum foil and place in a drying oven for two hours at 110°C. When the tube has cooled, reweigh and calculate the dry weight of your cells in the aliquot. This represents the dry weight of starting material for your chromatically grown cultures and your phycobiliprotein/chlorophyll determinations.

Phycobiliprotein determination

7. Centrifuge the fifth aliquot, with water washing as described above.
8. Extract the cell mass by blending with powdered dry ice (ca. 50 g) and adding the powder to ca. 10 mL of 0.05 M potassium phosphate buffer pH 6.8. Alternatively, the cells can be ruptured and extracted by passing the cells, suspended in the buffer, through a French pressure cell, or ultrasonicating the cells suspended in buffer.
9. Centrifuge the extract in a refrigerated centrifuge at 30,000 × g for 20 minutes. Recover the supernatant, make to a known volume with buffer, and record the absorbance, E, of the solution at the following wavelengths: 650 nm; 620 nm; 565 nm.

   (If you have a recording spectrophotometer determine the absorption spectrum of the extract between 700 nm and 350 nm.)
10. Calculate the concentrations of the three phycobiliproteins from the following equations (from Bryant et al. 1979).

$$\frac{\text{C-Phycocyanin}}{\text{mg mL}^{-1}} = \frac{E(620 \text{ nm}) - 0.72 \times E(650 \text{ nm})}{6.29}$$

$$\frac{\text{Allophycocyanin}}{\text{mg mL}^{-1}} = \frac{E(650 \text{ nm}) - 0.191 \times E(620 \text{ nm})}{5.79}$$

$$\frac{\text{C-phycoerythrin}}{\text{mg mL}^{-1}} = \frac{E(565 \text{ nm}) - 2.41(\text{Conc. CPC}) - 1.40 (\text{Conc. APC})}{13.02}$$

Chlorophyll a determination

11. Recover the pellet from the biliprotein preparation. Add a few mg of $MgCO_3$ and ca. 5 mL of acetone. Grind the suspension in a Ten-Broek homogenizer. Centrifuge the extract at top speed in a clinical centrifuge. Save the supernatant. Repeat the extraction until no further chlorophyll is extracted. Combine the extracts and record the volume.
    WORK IN FUME HOOD AND NO SMOKING.
12. Measure the absorbance of the solution at 665 nm.

    (If you have a recording spectrophotometer determine the absorption spectrum of the chlorophyll solution between 700 nm and 350 nm).
13. Calculate the chlorophyll *a* concentration from the equation.

$$\frac{\text{Chlorophyll } a}{\text{mg mL}^{-1}} = \frac{E(665 \text{ nm})}{74.5}$$

Alternatively, you can measure the chlorophyll *a* concentration by the fluorescence method described in Experiment 4B.

14. Express the results (now in mg mL^{-1}) of your phycobiliprotein and chlorophyll *a* determinations in terms of mg pigment g^{-1} dry weight cells.

Final Harvest

15. After three weeks, remove the green and red cellophane filters from the cultures. Do the cultures show any apparent color differences?
16. Examine the filaments under the microscope. Do you see any differences in cell shape?
17. Harvest the cells by centrifugation with a double wash at the centrifuge with distilled water and weigh the damp cell mass. Keep a small aliquot (ca. ⅛ cell mass) for dry weight determination. You will use the remaining aliquot for pigment determinations.
18. Determine the dry weight of the smaller aliquot as described in Step 6.
19. For each of the larger aliquots, determine the allophycocyanin, C-phycocyanin and C-phycoerythrin concentrations as described in Steps 8, 9, 10. Express your results in mg phycobiliprotein g^{-1} dry weight.
20. Determine the chlorophyll *a* content, according to Steps 11, 12, 13 in the residual pellet from the phycobiliprotein determination. Express your results as mg chlorophyll *a* g^{-1} dry weight.
21. Tabulate your results, with reference to absolute amounts of pigments, relative ratios, and ratios relative to allophycocyanin (set at 1).

Calculations

Light-grown experiment, red/green or white light. In these experiments you need to calculate the dry weight, total amounts, concentrations and ratios of pigments for the starting inoculum and the final harvests. You are also concerned with net synthesis under the different light regimes, so you must make allowance for the cell mass and pigments in your original inoculum. From your results calculate the net amount of pigment synthesis.

Questions

1. What effects do red light and green light have on the biosynthesis of each of the three phycobiliproteins? On chlorophyll *a*?
2. Do your results indicate that *Fremyella* shows type b chromatic adaptation as described in the introduction?
3. What effect does light quality have on the morphology of individual cells?
4. You expressed the phycobiliprotein ratios relative to allophycocyanin, which was designated as 1. Why?
5. What are possible sources of error?

REFERENCES

Beguin, S., G. Guglielini, R. Rippka & G. Cohen-Bazire (1985). Chromatic adaptation in a mutant of *Fremyella diplosiphon* incapable of phycoerythrin synthesis. *Biochimie* 67, 109–117.

Bennett, A. & L. Bogorad (1973). Complementary chromatic adaptation in a filamentous blue-green alga. *J. Cell. Biol.* 58:419–435.

Bogorad, L. (1975). Phycobiliproteins and complementary chromatic adaptation. *Ann. Rev. Plant Physiol.* 26:369–401.

Bryant, D. A., G. Guglielini, N. Tandeau de Marsac, A-M. Castets & G. Cohen-Bazire (1979). The structure of Cyanobacterial phycobilisomes: A model. *Arch. Microbiol.* 123, 113–127.

Gantt, E. (1980). Structure and function of phycobilisomes: light harvesting pigments complexes in red and blue-green algae. *Int. Rev. Cytol.* 66, 45–80.

Gantt, E. (1981). Phycobilisomes. *Ann. Rev. Plant Physiol.* 28, 355–377.

Glazer, A. (1982). Phycobilisomes: structure and dynamics. *Ann. Rev. Microbiol.* 36, 173–198.

Glazer, A. (1984). Phycobilisome. A macromolecular complex optimized for light energy transfer. *Biochim. Biophys. Acta.* 768, 29–51.

Ohki, K., E. Gantt, C. A. Lipschultz & M. C. Ernst (1985). Constant phycobilisome size in chromatically adapted cells of the Cyanobacterium *Tolypothrix tenuis,* and variation in *Nostoc* sp. *Plant Physiol.* 79, 943–948.

Rosinski, J., J. E. Hainfeld, M. Rigbi & H. W. Siegelman (1981). Phycobilisome ultrastructure and chromatic adaptation in *Fremyella diplosiphon. Ann. Bot.* 47, 1–12.

Scheer, H. (1982). Biliproteins. *Angew. Chem. Int. Edit.* 20, 241–261.

Tandeau de Marsac, N. (1983) Phycobilisomes and complementary chromatic adaptation in Cyanobacteria. *Bull. Inst. Pasteur* 81, 201–254

NOTES FOR INSTRUCTORS

Supplies for each pair of students

1. culture of *Fremyella diplosiphon* (UTEX 481) (approximately 600 mL of culture)
2. Ten Broeck homogenizer
3. 3 Fernbach flasks (or 2 L Erlenmeyer flasks) with 900 mL medium
4. red and green transparent cellophane, or any material that will provide broad band red light (with minimal or no green) and green light (with minimal or no red)
5. microscope and supplies
6. beakers (20, 50, 100, 250 mL); Erlenmeyers (20, 50, 125, 250 mL); measuring cylinders (10, 25, 50 mL); pipettes (1, 5, 10 mL), test-tubes
7. spatula
8. dry ice (ca 100g)

9. magnesium carbonate
10. 500 mL 0.05 M potassium phosphate buffer, pH 6.8.
11. 20 mL acetone
12. plastic gloves
13. aluminum foil
14. weighing bottles

Laboratory equipment

1. Clinical centrifuge with 15 and 40 mL tubes (4 each per student pair)
2. Refrigerated centrifuge with 40 mL tubes and 250 mL bottles (4 each per student pair)
3. Drying oven
4. Spectrophotometer and cuvettes
5. Analytical balance

Cultures

Since the inoculum from UTEX will be test-tube sized, the instructor should start working the culture up to appropriate size 9–12 weeks prior to this. An inoculum of 300–500 mL will be needed and should be ready in 6–8 weeks. About 3–4 weeks before the beginning of the experiment start to grow up 3 L of *Fremyella diplosiphon* in Kratz-Meyer D medium (see below), to a density of about 0.5–1 g fresh-weight cells per liter. This is done at 20°C in cool white fluorescent light, ca. 30–50 μE m^{-1} sec^{-1}. An inoculum of 300–500 mL per 3 L will produce an appropriate density in the 3–4 weeks. This 3 L culture is the basic inoculum and will provide sufficient cells for five pairs of students.

White, green and red light cultures should all be grown at equal intensities of approximately 30 μE m^{-2} sec^{-1}. Ensure that the green and red light cultures are completely surrounded by appropriate cellophane.

If space for cultures is a problem, the instructor may wish to grow up the white, red and green light final cultures in 12 L carboys (10 L of medium) aerated with filtered air. The inocula are scaled up appropriately, and a 10 L culture can then be divided between the class.

It is instructive to record the transmission-absorbance characteristics of the green and red cellophane. This is easily done with the recording spectrophotometer.

Fremyella diplosiphon is obtainable as strain UTEX 481 from the University of Texas Culture collection of Algae, Austin Texas. *Tolypothrix*

tenuis may also be used instead. The growth medium is Kratz-Meyers D
medium (American Journal of Botany 42, 282–287, 1955):

K_2HPO_4	1 g
$NaNO_3$	1 g
$MgSO_4 \cdot 7H_2O$	150 mg
$CaNO_3 \cdot 4H_2O$	10 mg
Na_2EDTA	50 mg
$Fe_2(SO_4)_3 \cdot 6H_2O$	4 mg
Trace Elements	1 mL
Distilled H_2O	To 1000 mL

Trace Elements Mix

$ZnSO_4 \cdot 7H_2O$	8.82 g
$MnCl_2 \cdot 4H_2O$	1.44 g
$CuSO_4 \cdot 5H_2O$	1.57 g
$CoCl_2 \cdot 6H_2O$	0.40 g
H_2O	To 1000 mL

Comments

The experiment described here is set up for student pairs. It is suggested
that each pair be responsible for all the calculations for one light regime.

Extraction of the biliproteins (Step 8) can cause a problem. You may
wish to experiment beforehand to determine the best method.

14

Extraction and assay of ribulose-1,5-bisphosphate carboxylase from *Fucus*

J. N. Keen and L. V. Evans

Department of Plant Sciences, University of Leeds, Leeds LS2 9JT, United Kingdom

FOR THE STUDENT

Introduction

In photoautotrophic organisms, light energy is converted into chemical potential (NADPH and ATP), most of which is then used to convert inorganic carbon into diverse reduced carbon compounds by the photosynthetic carbon reduction cycle.

The primary step in this photosynthetic fixation of carbon is catalyzed by the enzyme ribulose-1,5-bisphosphate carboxylase/oxygenase (Rubisco; EC 4.1.1.39):

$$RuBP + CO_2 + H_2O \xrightarrow{\text{Rubisco}} 2 \times 3\text{-Phosphoglycerate}$$

In higher plants, green, red, and brown algae, Rubisco is a multimeric enzyme comprised of eight large (MW 53,000–55,000) and eight small (MW 12,000–15,000) subunits. The large subunits contain activator-bind-

Abbreviations:

dpm	:	disintegrations per minute
DTT	:	dithiothreitol
EDTA-Na$_2$:		ethylenediaminetetra-acetic acid-disodium salt
PEPCK	:	phosphoenolpyruvate carboxykinase
PVPP	:	polyvinylpolypyrrolidone
Rubisco	:	ribulose-1, 5-bisphosphate carboxylase/oxygenase
RuBP	:	ribulose-1, 5-bisphosphate

ing, substrate-binding and catalytic sites and are encoded by the chloroplast genome and synthesized in the chloroplast. In green plants, the small subunits are encoded by the nuclear genome and synthesized in the cytoplasm as 20kDa precursor polypeptides, which are cleaved as they enter the chloroplast. Here they associate with the large subunits to form active holoenzyme. The small subunits are believed to have a regulatory function. In red algae, and probably in brown algae, the small subunit appears to be synthesized within the chloroplast.

Rubisco is a bifunctional enzyme and can also catalyze the oxygenation of ribulose-1,5-bisphosphate (RuBP) in the process of photorespiration:

$$RuBP + O_2 \rightarrow \text{Phosphoglycolate} + \text{3-Phosphoglycerate}$$

The rates of the two reactions are controlled by the relative amounts of carbon dioxide and oxygen. Usually, there is sufficient carbon dioxide available and the carboxylation reaction is favored.

For a review of the structure and catalytic properties of Rubisco, see Jensen & Bahr (1977), Lorimer (1981) or Kerby & Raven (1985).

In brown seaweeds, there is another pathway of carbon fixation, which is not dependent on light. This light-independent fixation is catalyzed by the enzyme phosphoenolpyruvate carboxykinase (PEPCK; EC 4.1.1.32) and can account for as much as 20% of the carbon fixation, e.g., in *Laminaria* (Küppers & Kremer 1978):

$$\text{Phosphoenolpyruvate} + ADP + CO_2 \rightarrow \text{Oxaloacetate} + ATP$$

In the present experiment, a crude enzyme preparation will be extracted from the brown alga *Fucus,* by a procedure modified from that of Kerby & Evans (1983). A large proportion of the protein in this extract will be Rubisco, the carbon fixation activity of which will be determined at different temperatures, by a method based on that of Kerby & Evans (1981). The temperature optimum *(in vitro)* of RuBP carboxylation by the enzyme will be estimated, this will be related to the normal growing temperature of *Fucus,* and the effect on growth will be discussed.

Procedure

Note: This experiment involves the use of radioactive carbon. Please take the necessary precautions when working with radioisotopes.

Extraction

1. You are supplied with a tube containing tissue prepared by grinding *Fucus* apices in liquid nitrogen. Thaw the frozen powder in 50 mL of

ice-cold extraction buffer in a 250 mL beaker, as follows. Empty the vial of frozen powder into the buffer and break up the pellet with a spatula. It is important that this be done quickly so that lumps are removed before the tissue begins to thaw. Once thawed, the alginate in the tissue swells, making any remaining lumps very sticky, and it is not possible to disperse them, leading to low yields of enzyme.

The extraction buffer contains EDTA, DTT, ascorbate, Tween 80 and PVPP. How do these compounds help in the extraction of active enzymes from *Fucus*?

2. Add a magnetic stir bar to the extract. Place the beaker in an ice-bath on a magnetic stirrer and stir quite vigorously for 10 min to allow the tissue to thaw completely.

3. Place a square of muslin (20 cm × 20 cm, four layers thick) onto a funnel and pour the homogenate into it. The homogenate will be too viscous to drip through the muslin (what causes this viscosity?), so CAREFULLY squeeze the liquid through, collecting it in the measuring cylinder. Retrieve the stir bar. Record the volume of the filtrate and pour it into a cold ($-20°C$) 100 mL beaker.

4. Add sufficient $CaCl_2 \cdot 2H_2O$ to give a final concentration of 0.5% (w/v) (e.g., 0.2 g to 40 mL extract).

5. Wash the stir bar with distilled water and place it in the beaker. Stir the extract vigorously on a magnetic stirrer for 10 min.

What effect does the addition of $CaCl_2$ have on the extract? What is the chemical basis of this effect?

6. Transfer the extract to a 50 mL centrifuge tube and balance carefully against that of another group (by adding sand).

7. Centrifuge at 35,000 × g, for 30 min at 2°C to pellet the precipitate.

8. Retain the supernatant, which represents a crude enzyme preparation. Store on ice until required.

Rubisco assay

1. Sixteen numbered scintillation vials and 16 caps are provided. Record which numbers you use as tests or controls and the incubation temperature of each. This is to help the demonstrator who will dry the assays down, count them, and return the results. Do NOT write on the vials–this may prevent counting. IF YOU DO NOT KEEP CAREFUL RECORDS IT MAY NOT BE POSSIBLE TO INTERPRET THE RESULTS.

2. To each vial, add 100 μL of reaction mixture, containing $NaH^{14}CO_3$.

3. Add 50 μL enzyme extract to each vial and cap. Swirl gently to mix.

4. Incubate 4 vials at each temperature (10°C, 20°C, 30°C, and 40°C) for 1 h, to allow temperature equilibration and activation of the enzyme.

5. To two vials from each set of four, add 100 μL RuBP-Na$_4$ solution. To the other two vials (controls) from each set, add 100 μL control buffer.
6. Incubate for exactly 15 min more at the various temperatures.
7. Terminate the reactions IN A FUME CUPBOARD (why?) by adding 0.5 mL 6 M acetic acid to each vial. Leave the vials uncapped in the fume cupboard.

   Stages 8–11 will be carried out for you.
8. Dry the vials down on a hot plate at 50°C to drive off unused $^{14}CO_2$. This procedure is best carried out overnight to ensure complete drying.
9. Add 0.5 mL water to each vial. Swirl to dissolve the residue.
10. Add 10 mL scintillation cocktail. Replace caps and shake.
11. Count the radioactivity in a scintillation counter to determine the amount of $H^{14}CO_3^-$ incorporated.
12. Calculate means of duplicates and subtract controls from test reactions for each temperature.
13. Plot dpm ^{14}C incorporated against temperature and estimate the temperature at which Rubisco activity is optimal *in vitro*. (Why might this not be the same for the enzyme *in vivo?*)

    How does this compare with the temperature at which *Fucus* normally grows? What effect might this have on the growth of the organism?

    What compounds are likely to become labelled in this experiment? How might you test this?

    Why is the extraction of active enzymes from brown seaweeds difficult, and how are these problems overcome?

REFERENCES

Jensen, R. G. & J. T. Bahr (1977). Ribulose 1,5-bisphosphate carboxylase-oxygenase. *Ann. Rev. Plant Physiol.*, 28, 379–400.

Kerby, N. W. & L. V. Evans (1981). Pyrenoid protein from the brown alga *Pilayella littoralis. Planta*, 141, 469–75.

Kerby, N. W. & L. V. Evans (1983). Phosphoenolpyruvate carboxykinase activity in *Ascophyllum nodosum* (Phaeophyceae). *J. Phycol.* 19, 1–3.

Kerby, N. W. & J. A. Raven (1985). Transport and fixation of inorganic carbon by marine algae. *Adv. Bot. Res.*, 11, 71–123.

Küppers, U. & B. P. Kremer (1978). Longitudinal profiles of carbon dioxide fixation capacities in marine macroalgae. *Plant Physiol.*, 62, 49–53.

Lorimer, G. H. (1981). The carboxylation and oxygenation of ribulose 1,5-bisphosphate: The primary events in photosynthesis and photorespiration. *Ann. Rev. Plant Physiol.* 32, 49–383.

NOTES TO INSTRUCTORS

Materials

Although we use *Fucus serratus,* the extraction procedure and carbon fix-ation assay can also be used for other *Fucus* spp. and *Ascophyllum.* It is probable that this experiment will also work using other brown algae, but our investigations have been restricted to fucoids.

Preparation of tissue

Wash 5.0 g of non-fertile *Fucus* apices in distilled water and blot dry. Cut into small pieces, approximately 0.5 cm². Grind the tissue to a fine powder in liquid nitrogen.

Using a Freezer/Mill, tissue grinding is fast and easy, a grind of several minutes at fairly high frequency reducing the pieces of tissue to a fine powder, suitable for enzyme extraction. If use of a Freezer/Mill is not possible, the tissue can be ground by hand in liquid nitrogen, using a mor-tar and pestle, but this takes much longer and is extremely hard work, especially if a large number of samples are to be prepared. In addition, loss of material from the mortar, due to the boiling of the nitrogen, is almost unavoidable.

The frozen powder produced (by either method) should be transferred to 50 mL plastic centrifuge tubes and stored at $-20°C$ in 5 g amounts. The powder is best used within two weeks; after that it becomes difficult to break the pellet up. Although the ground tissue can be used immedi-ately, it may prove difficult to thaw it in buffer, as its temperature is so low $(-196°C)$; thus it is recommended that the tissue be prepared in advance and stored at $-20°C$.

Special equipment

1. Freezer/Mill and accessories (grinding vials, extractor and vial opener), (e.g., Spex 6700 available from Spex Industries Inc., Metuchen, N.J., U.S.A.) or porcelain mortars and pestles.
2. Liquid nitrogen.
3. Refrigerated centrifuge and 8×50 mL rotor, both cooled to 2°C.
4. Scintillation counter.

Equipment per group of students

1. 250 mL glass beaker, cooled to $-20°C$ (1).
2. 100 mL glass beaker, cooled to $-20°C$ (1).

3. 1 L plastic beaker (for ice-bath) (1).
4. 100 mL measuring cylinder (1).
5. 10 cm diameter funnel (1).
6. 50 mL plastic centrifuge tube (1).
7. 16 glass scintillation vials, caps, which are numbered (16).
8. large spatula (1).
9. small spatula and weighing boat (1).
10. 20 cm × 20 cm muslin (or cheesecloth) square (four layers thick).
11. magnetic stirrer and 4 cm stir bar.
12. ice bucket.

Class equipment

1. incubators at 10°C, 20°C, 30°C and 40°C.
2. hot plate at 50°C in fume cupboard.
3. automatic pipettes and tips to deliver 50 μL, 100 μL, and 500 μL.
4. Pasteur pipettes.
5. acid-washed sand, for balancing centrifuge tubes.
6. graph paper (in cm).

Chemicals required

1. Freshly made extraction buffer (50 mL per group). Store on ice.
   0.1 M Tris-HCl (pH 8.0) containing:
   5 mM Mg acetate·4H$_2$O (or MgCl$_2$·6H$_2$O)
   2 mM EDTA-Na$_2$
   5 mM DTT
   10 mM L-ascorbate-Na (or L-ascorbic acid)
   2 g L^{-1} Tween 80
   100 g L^{-1} PVPP
   The PVPP (supplied as a powder) must be treated before use. It is washed for 10 min in 10% HCl, then washed with distilled water, neutralized with 10% KOH (to pH 7 using pH meter) and given a final wash with distilled water (Loomis 1974). It is then hydrated in 0.1 M Tris-HCl, pH 8.0 (100 g PVPP in 40 mL buffer). It is stored at 4°C (indefinitely) and used in this form. PVPP is an insoluble polymer of polyvinylpyrrolidone and hence will not dissolve. Stir the solution to disperse it before dividing into 50 mL aliquots.
2. CaCl$_2$·2H$_2$O (approximately 0.2 g per group).
3. Reaction mixture:
   Stage 1 (Store at room temperature).
   0.1 M Tris-HCl (pH 8.0) containing:

20 mM Mg acetate·4H$_2$O
50 mM NaHCO$_3$
Stage 2 (2 mL per group). Make fresh.
Add to stage 1 buffer:
10 mM DTT
20 μCi mL^{-1} NaH^{14}CO$_3$
4. Control buffer (1mL per group). Store at room temperature.
   0.1 M Tris-HCl (pH 8.0)
5. Ribulose-1,5-bisphosphate-tetrasodium salt (RuBP-Na$_4$) (1 mL per group).
   3 mM RuBP-Na$_4$ in control buffer
   Make up immediately before use and store on ice.
6. 6 M acetic acid (10 mL per group).
7. Scintillation cocktail (200 mL per group).
   8 g L^{-1} 2,5-diphenyloxazole (PPO) in toluene:Triton X-100 (2:1).

RuBP-Na$_4$ is available from Sigma Chemical Company as a 70% pure grade which is suitable. Triton X-100, toluene and PPO should be of scintillation grade. All other reagents should be of analytical grade where possible.

Comments

The extraction buffer must be freshly-made on the day of use and stored on ice until required. L-ascorbic acid can be used instead of sodium salt, and magnesium chloride can replace the magnesium acetate (molar for molar replacement).

The homogenate produced is very viscous, due to the presence of large amounts of alginate, necessitating filtration by squeezing through muslin.

Treatment with calcium chloride causes precipitation of much of the alginate as an insoluble Ca-alginate gel, due to cross-linking of guluronic acid-rich alginate (GG blocks) by the calcium. This gel can then be removed by centrifugation to leave a crude protein extract of much lower viscosity.

The extract contains Rubisco as a major protein component, its subunits being the major bands on sodium dodecyl sulphate-polyacrylamide gel electrophoresis (SDS-PAGE).

It is recommended that the instructor provide prenumbered vials because of possible nonadherence to the instructions concerned with setting up the Rubisco assays.

Assays are carried out at four different temperatures to give an idea of the temperature optimum of the enzyme *in vitro*. This may not be the same for the enzyme *in vivo,* since concentrations of activators and substrates may be different (thus affecting enzyme activity), so data should be interpreted with caution. Duplicate assays are performed to allow identification of any errors, with or without the substrate, RuBP. The controls (without RuBP) will give an indication of the background radioactivity, which may include any unused $NaH^{14}CO_3$ not removed in the drying step or any fixation by PEPCK using endogenous substrate. In practice, the level of background radioactivity is very low, indicating that light-independent fixation is absent and that all unused $^{14}CO_2$ is driven off.

Normally, in this (^{14}C-fixation) assay, initial incubation times are short, usually 15 min, which is sufficient to allow the enzyme to reach its optimal temperature and become activated by magnesium and bicarbonate. Here, the enzyme must reach temperatures far from its optimum and it is considered that a longer period of equilibration and activation is necessary.

Reactions should be terminated in a fume cupboard, since radioactive carbon dioxide is released.

About $2\mu Ci$ $NaH^{14}CO_3$ is sufficient to give good incorporation of counts, but the actual specific activity is not critical. The $NaH^{14}CO_3$ is diluted with unlabelled bicarbonate here to allow amounts of fixed carbon (μmol) to be calculated if required. However, it is not necessary for determining the optimal temperature of the enzyme.

The graph of dpm $H^{14}CO_3$ incorporated into acid–stable products versus temperature should show the typical normal distribution curve of enzyme reactions, the temperature optimum peak appearing around 25–30°C. At temperatures on either side of this optimum, the reaction rate is slower, as a result of a lower metabolic rate at low temperatures or partial protein denaturation at higher temperatures. If rates are not reduced at higher temperatures, it may be that the initial incubation period was not long enough for this to occur.

The results should show that the temperatures in the natural habitat are suboptimal and that the enzyme can achieve much higher rates than it would in the environment (sea temperature is fairly constant at 5°–15°C). This would suggest that *Fucus* growth rates could be much higher, with temperature being a limiting factor by preventing enzyme reactions from reaching their maximal rates. It should be noted that growth rates can be significantly higher during brief periods of exposure to air (Strømgren 1977), when temperatures and irradiance are likely to be higher than during submergence, thus supporting the view that Rubisco can function at a higher rate in optimal conditions. However, although not normally optimal, enzyme rates are sufficiently high to allow growth at such sea temperatures. It is probable that catabolic reactions, such as respiration, are also operating at lower than maximal rates, thus enabling growth to occur.

The problems in the extraction of enzymes from brown algae are four-fold (Marsden, Callow & Evans 1981). First, brown algae contain large amounts of polyphenols, which are oxidized to enzyme-inhibitory tannins on extraction. Second, brown algal mucilages, such as alginate, can bind to and inhibit certain enzymes. They also create problems of low pH and high viscosity. Third, brown algal tissue is very tough, rendering conventional homogenization techniques ineffective. Grinding in liquid nitrogen is the most effective method, producing a high degree of cell breakage, preventing alginate from swelling and restricting the oxidation of polyphenols. Finally, brown algae contain only low amounts of protein, making the use of fairly large amounts of material necessary.

Various protective agents should be included in the extraction buffer to assist in the extraction of active enzymes, e.g., PVPP to complex polyphenols, EDTA as a heavy metal chelator, DTT as a protein SH-group protector, ascorbic acid as an anti-oxidant and Tween 80 as a membrane solubilizer.

Additional References

Loomis, W. D. (1974). Overcoming problems of phenolics and quinones in the isolation of plant enzymes and organelles. *Methods Enzymol.*, 31, 528–544.

Marsden, W. J. N., J. A. Callow & L. V. Evans (1981). A novel and comprehensive approach to the extraction of enzymes from brown algae, and their separation by polyacrylamide gel electrophoresis. *Mar. Biol. Lett.*, 2, 353–362.

Strømgren, T. (1977). Length growth rates of three species of intertidal Fucales during exposure to air. *Oikos*, 29, 245–249.

15

Thin-layer chromatographic analysis of polyols (alditols)

Bruno P. Kremer

Universität zu Köln, Institüt für Naturwissenschaften und ihre Didaktik, Abteilung für Biologie, 5000 Köln 41, Federal Republic of Germany

FOR THE STUDENT

Introduction

Most green plants accumulate their soluble, low-molecular weight photosynthates in the form of free mono- or disaccharides (and even oligosaccharides). On the other hand, many plants also contain certain low-molecular weight carbohydrates derived from aldoses and ketoses by reduction of their carbonyl group to a hydroxyl group. Each C-atom of such compounds thus carries an OH-group (Fig. 15.1), which is why they are generally referred to as sugar alcohols, alditols, or (preferably) polyols. Plants with polyols gain one mol reduction equivalent more upon glycolytic degradation of these compounds than with the catabolism of an ordinary sugar. The reduction of a photosynthate beyond the level of an ordinary aldose or ketose therefore appears to be an efficient way to conserve reducing power. However, as you will see from this experiment, the distribution of the various polyols is very scattered and they are by no means ubiquitous.

Although there are also cyclic compounds such as the inositols, conventionally only the compounds with an open C-chain are understood as polyols (Fig. 15.1). Whether glycerol should be included in this group of

150

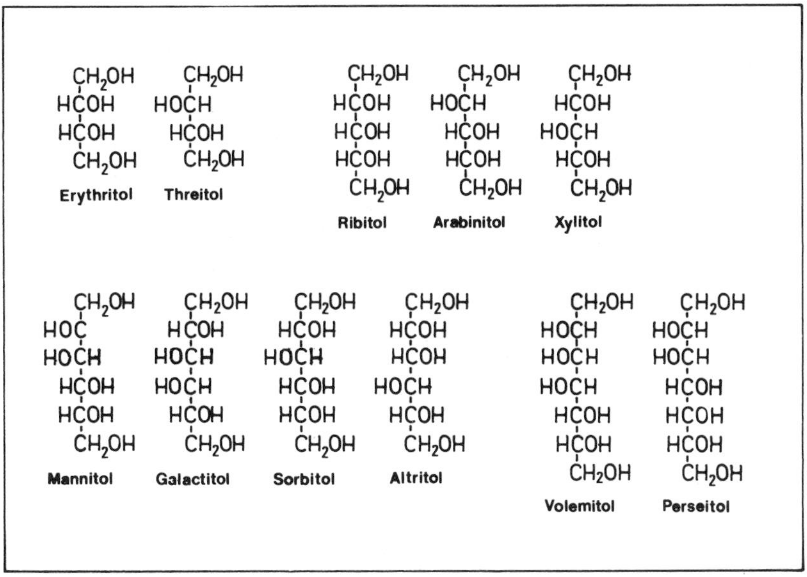

Fig. 15.1. Chemical structure of some naturally occurring polyols.

natural products has been debated. Glycerol forms an important structural component of lipids and, moreover, occurs as free photosynthate in a variety of unicellular algae. Only two of the four theoretically possible tetritols (= C_4-polyols; erythritol and threitol) have been found in nature. Similarly, the pentitols (C_5-polyols) have a very scattered distribution. They comprise compounds such as arabinitol (= lyxitol), ribitol (= adonitol), and xylitol. Ten stereoisomeric hexitols (= C_6-polyols) are theoretically possible. Only six of them have been traced among natural products, i.e., allitol, iditol, altritol, mannitol, sorbitol (= glucitol, gulitol), and galactitol (= dulcitol). Only the latter three are more frequently encountered. Polyols with chain lengths exceeding six C-atoms (e.g., volemitol, perseitol) are extremely rare.

Marine and freshwater algae include a variety of taxa which accumulate photosynthates as polyols; hence they represent interesting and promising candidates for further analysis. Since relatively few species have yet been studied in detail, new discoveries are not unlikely. In this experiment we therefore investigate (1) which polyols are found among algae, (2) which common algae contain which polyol(s), and (3) whether or not the resulting pattern of polyol distribution is of taxonomic significance.

Procedure

Extraction. Because all polyols are water-soluble and usually available in the algae in large quantities (if present at all), the first step of the analysis includes the preparation of a plant extract. Homogenize 1–2 g fresh weight of any available algal species or about 0.5 g freeze-dried material by grinding in a mortar with quartz sand. The addition of some liquid N_2 is very helpful particularly in the case of the leathery brown algae. Extract the algal powder for 10–15 min with 5–10 mL 30% ethanol. The homogenate is then cleared by filtration (through ordinary filter paper) or by centrifugation (500 × g, 5 min). The resulting liquid can be used for chromatographic separation without further purification.

Chromatography. Thin-layer plates are charged with about 3–5 µL of the extracts as well as with reference solutions of authentic polyols, by means of micropipettes or glass capillaries. Apply all extracts and solutions as small (ca. 10 × 22 mm) bands, since this will encourage a more distinct separation during the development of the thin-layer plate. Avoid damaging or scratching of the thin-layer by the tip of the pipettes! If you plan to test the extract(s) for ketoses and aldoses, two additional thin-layer plates must be prepared.

For development, conventional 4-L glass tanks are used. One run (one-dimensional development of the chromatogram) takes about 4–5 h. The run is stopped when the solvent front reaches the upper edge of the thin-layer plate. Next, allow the plates to dry in a fume hood for 30–60 min.

Detection. Polyols are detected by spraying the plates in a fume hood (1) with a solution of sodium metaperiodate and (2) with benzidine reagent. CAUTION! Benzidine reagent is very toxic and a suspected carcinogen! Avoid inhalation and contact with the skin! Spray solution (1), wait for 1–2 min, and then spray the benzidine reagent. This detection procedure reliably demonstrates amounts of polyols even in the range of 1 µg. The polyols show up as yellow-white spots on a dark blue background at room temperature about 1–2 min after spraying the benzidine reagent (Fig. 15.2).

The contrast between the detected spots and their background begins to vanish after 10–15 min. There is no reliable procedure of stabilization or preservation. Hence the results of the chromatographic separation are immediately transferred by marking on a transparent sheet (e.g., overhead projector sheet) spread over the plates. It is not necessary to dry the plates completely or to handle them with gloves.

Thin-layer chromatograms prepared by this technique can also be ana-

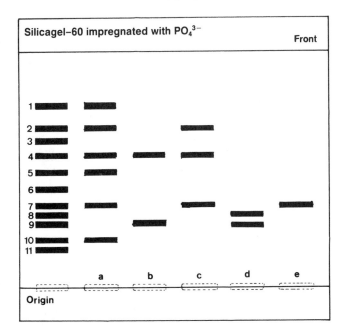

Fig. 15.2. Schematic chromatogram of polyols on silicagel-60 (impregnated with phosphate). 1 = glycerol, 2 = erythritol, 3 = threitol, 4 = ribitol, 5 = arabinitol, 6 = xylitol, 7 = mannitol, 8 = galactitol, 9 = sorbitol, 10 = volemitol, 11 = perseitol. Channels a–e are examples of algal extracts containing polyols, namely a = *Trentepohlia* sp., b = *Prasiola* sp., c = *Pleurococcus* sp., d = *Bostrychia* sp., e = *Fucus* sp.

lyzed for the presence of free mono- or disaccarides using different detection reagents. For the detection of aldoses, an aniline-phthalate spray is used. The detection of ketoses is achieved by spraying with urea-HCl reagent. In both cases, the thin-layer plates are heated for 3–5 min at 110°C upon spraying. Aldopentoses show up as brick-red spots, aldohexoses as browns, aldoheptoses as blue-green. Ketoses are recognized by their blue-gray colored spots. The presence of polyols in an extract does not interfere with the detection of aldoses or ketoses and vice versa.

Questions

1. Which species contain one or more polyols? Are there more species with one or with several polyols?
2. Which polyols are most frequently detected in your samples? Which compounds did you find in only one or few samples?

3. Provided you found a variety of algae containing two or three different polyols: Can you trace some common features in the autecology of the species?
4. Is it possible to delimit patterns of distribution? Does the occurrence of polyols correspond to algal taxonomy?
5. What about simultaneous occurrence of polyols, aldoses and ketoses? Are there any compounds that are mutually exclusive?
6. How could you define the terms 'occurrence' and 'accumulation'?
7. What environmental factors might the absolute amounts of polyols depend on?
8. Which other factors might control the accumulation? Try to design a research project that could add further information.

REFERENCES

Bielesky, R. L. (1983). Sugar alcohols. In, F. A. Loewus & W. Tanner (eds.) *Encyclopedia of Plant Physiology, Vol. 13A, Plant Carbohydrates*. Springer, New York, pp. 158–192.

Chudek, J. A., R. Foster, I. R. Davison & R. H. Reed (1984). Altritol in the brown alga *Himanthalia elongata*. *Phytochemistry* 23, 1081–1082.

Kremer, B. P. (1978). Improved method for the thin-layer chromatographic identification of alditols. *J. Chromat.* 166, 335–337.

Kremer, B. P. (1981) Carbon metabolism. In, C. S. Lobban & M. J. Wynne (eds.) *The Biology of Seaweeds*. Blackwell Scientific Publications, Oxford, pp. 493-533.

Kremer, B. P. & G. O. Kirst (1982). Biosynthesis of photosynthates and taxonomy of algae. *Z. Naturf.* 37c, 761–771.

NOTES FOR INSTRUCTORS
Equipment and supplies

1. balance
2. mortar and pestle (for each student)
3. oven (to give 110°C)
4. centrifuge
5. centrifuge tubes (10 mL) according to sample numbers
6. transparent sheet (overhead projector sheet), marker
7. 4-L glass tanks for development of 20 × 20 cm thin-layer plates
8. spraying flasks
9. quartz sand (1 kg)
10. liquid N_2 (optional)
11. 30% ethanol
12. 5μL micropipets

Chromatography supplies. Thin-layer plates (glass plates 20 × 20 cm) coated with silica gel 60 (e.g., Merck article No. 5715). Use commercially available ready-for-use plates. For this experiment, these plates must be impregnated with sodium phosphate. Accordingly, about 30 mL of a 0.5 M solution of $NaH_2PO_4 \cdot 2\ H_2O$ (78 g L^{-1}) in 50% methanol are sprayed on the thin-layer or the whole plate is briefly submersed in an equivalent volume of that solution. The plate is then stored horizontally for 30 min at room temperature and subsequently activated at 110°C for 60 min. Thin-layer plates treated by this procedure can be stored for a long time (months or years) without any loss of capacity.

Reference solutions of authentic polyols (and sugars) (0.5–1% in 30% ethanol; add 1 drop pyridine/5 mL to prevent microbial degradation) should include the following compounds:

–glycerol	–ribose	–glucose
–threitol	–mannitol	–fructose
–arabinitol	–sorbitol	–galactose
–ribitol	–galactitol	–sucrose

Solvent: iso-propanol-acetone-0.2 M lactic acid 40:40:20 (100 mL per tank). This mixture must be prepared daily and should be ready 1–2 h before the development of the thin-layer plates starts.

Detection reagents

Polyols: (1) 0.1% aqueous solution of sodium metaperiodate
(2) benzidine reagent: dissolve 1.8 g benzidine in 50 mL 96% ethanol, add 50 mL H_2O after complete dissolution, then 20 mL acetone and 10 mL 0.2 N HCl.
Aldoses: Weigh 0.93 g aniline (liquid) from a pipette into a vial. Dissolve aniline and 1.66 g phthalic acid in 100 mL n-butanol previously saturated with H_2O.
Ketoses: Dissolve 5 g urea in 20 mL 2 N HCl and add 100 mL 96% ethanol

Scheduling

Plant extracts, charged and developed thin-layer plates (before detection of the separated compounds) can be stored at room temperature. Hence, the major steps of this experiment could be performed on subsequent days.

– extraction of samples:	1h
– loading of plates:	1h

– development of plates:	5h
– detection/documentation:	1h
– discussion of results:	1h

Comments

In order to evaluate a differential pattern of polyol distribution in marine and freshwater algae, a selection of at least 10 species should be analyzed. Since the extraction does not involve any major technical handling, each student of a group could process two to three different algal samples within reasonable time. The selection of species should include some brown algae, all of which contain large amounts of mannitol. The European brown algal species *Himanthalia elongata* and *Pelvetia canaliculata* both accumulate two different polyols. Red algae usually do not photosynthesize polyols, with the exception of the genus *Bostrychia* (with sorbitol and galactitol). Green algae normally contain a wide variety of sugars, but no polyols. However, almost all species living aerophytically (e.g., *Cephaleuros, Pleurococcus, Prasiola, Stichococcus, Trentepohlia*) have been found to belong to the polyol plants. Hence, the sample selection should consider such species. *Dunaliella* spp. accumulate free glycerol. Many further genera and species await detailed examination.

When the laboratory is equipped with the machinery for gas-liquid chromatography (GLC) or even high-performance liquid chromatography (HPLC), the identification of polyols simultaneously allows for their quantification in the respective sample. If such equipment is not available, but a quantitative determination of polyol contents is desired, the classic periodate oxidation procedure according to Lewis & Harley (*New Phytol.* 64, 225–32 [1965]) could be used (see Exp. 25).

16

Chromatographic identification of component sugars in the extracellular polysaccharide of *Porphyridium*

David J. Chapman

Dept. of Biology, University of California, Los Angeles, CA 90024, USA

FOR THE STUDENT

Introduction

A number of unicellular rhodophytes produce copious amounts of extracellular polysaccharides. This is particularly noticeable in older cultures where the medium can become very viscous. Genera and species that have been examined include *Rhodella maculata* (Evans et al., 1974); *Porphyridium aerugineum* (Ramus 1972, 1973, 1986; Percival and Foyle 1979; von Witsch et al 1983); and *Porphyridium cruentum* (Medcalf et al. 1975; Heaney-Kieras and Chapman 1976; Heaney-Kieras et al. 1977; Percival and Foyle 1979).

The purpose of this experiment is to isolate the extracellular polysaccharide from the medium and do a preliminary characterization of the component monosaccharides. You will ignore the polypeptide component that is covalently bound to the polysaccharide.

Procedure

Preparation of polysaccharide. You will be provided with a culture of *Porphyridium cruentum* (ca 500 mL). Note the viscosity of the culture.

157

Examine the cells, with their surrounding mucilage, under the microscope.

1. Centrifuge the culture for 30 min at 10,000 × g at 15°C.
2. Recover the supernatant, ensuring that all the cells have been pelleted. You may discard the pellet.
3. Take a known aliquot of the supernatant (ca 300–400 mL) and dialyse against running tap water for 48 h. After this, transfer the solution to fresh dialysis tubing and dialyse for 24 h against two changes of distilled water.

   The instructor will demonstrate the preparation of dialysis bags with presoaking and tying the ends of the tubing. It is recommended that you use about 60–70 mL supernatant per tube (diameter ⅝″) and dialyse against large volumes of distilled H_2O (ca 5 L). The instructor will carry out the dialysis.

Hydrolysis of polysaccharide

4. At the end of three days dialysis, recover the supernatant from the dialysis tubing. Stir in dry NaCl to bring the solution to 0.05 M NaCl. Follow this by the slow addition of sufficient cetylpyridinium chloride (CPC) to make 1% CPC solution. Bring the solution to 40°C over a warm plate. The polysaccharide will precipitate out.
5. Recover the precipitate by centrifugation at 20,000 × g for 30 min. Wash the precipitate first with ethanol; then ethanol : diethyl ether (1:1 v/v); and finally diethyl ether. Dry the polysaccharide in a desiccator under vacuum. CAUTION: USE FUME HOOD.
6. Weigh the polysaccharide.
7. Take ca 20 mg of polysaccharide and add 10 mL of 1 N H_2SO_4.
8. Hydrolyse at 100°C for 2 h in a sealed, evacuated, vacuum hydrolysis tube.
9. Let the contents cool, and neutralize by the addition of $BaCO_3$. CAUTION: THE REACTION WILL FOAM.
10. Remove the $BaSO_4$ and any excess $BaCO_3$ by centrifugation (10 min at 10,000 × g) or by passage through a fine porosity sinter glass funnel under vacuum.

Chromatography of hydrolysate

11. You will be provided with thin layer plates of silica gel G that have been impregnated with 0.02 M sodium acetate, or regular unimpregnated plates.
12. Spot 30μL of the hydrolysate onto the silica gel together with 20 μL of each of the four standard monosaccharide solutions adjacent to the hydrolysate solution (see Figure 9.1).

CAUTION: a) Don't break the silica gel surface. Let capillary action "draw out" the drops.

b) Let the spot dry partially (fan blower helps) before adding additional spots.

c) Place your spots about 2.5 cm from the bottom of the plate in a uniform line and separated from each other by about 2 cm.

13. Allow the spots to dry, then place the plate in a chromatographic chamber (lined with filter paper as in Experiment 9) with one of the following three solvents (120 mL solvent per chamber).

    Chloroform : methanol 60:40 v/v

    Ethyl acetate : methanol : acetic acid : H_2O 60:15:15:10 v/v/v/

    n-butanol : acetic acid: diethyl ether : H_2O 45:30:15:5 v/v/v/v

    (Use the last system for the unimpregnated plate and the first two solvent systems for the impregnated plate)

14. Let the plates develop to within about 3 cm from the top. Remove from the chamber and allow to dry (CAUTION: DRY IN A FUME HOOD).

15. Drying can be facilitated by blowing with an air current, or placing in a warm oven (ca 50°C).

16. Spray the dried plate with one of the following visualizing reagents.
    Naphthol-resorcinol
    Anisaldehyde-sulfuric acid
    Anisidine-phthalate
    CAUTION: SPRAYING MUST BE DONE IN A FUME HOOD.

    a) Place the plate flat down on a large sheet of paper.

    b) Spray the plate uniformly with the reagent by means of a solution atomizer or propellant system. Spray the plate to dampness: Too much reagent and the solution will run. The reagent spray must be allowed to soak into the plate. Repeat the spray until the adsorbent is uniformly moist.
    CAUTION: WEAR DISPOSABLE GLOVES.

17. Heat the plate at 100°C for 5–10 min to visualize the carbohydrates. Visualization time can vary. Watch the plate until the carbohydrate spots are visible. Avoid extensive heating, which can produce background coloration.

18. Remove the plate and allow to cool.

Questions

1. What sugars can you identify from your plate?

2. How many other unidentified sugars are there? From your knowledge of the literature, do you have any suggestions of what these other carbohydrates are?

3. What are possible sources of contaminants?

REFERENCES

Evans, L. V., M. E. Callow, E. Percival, & V. Fareerd (1974). Studies on the synthesis and composition of extracellular mucilage in the unicellular red alga *Rhodella. J. Cell Sci.* 16, 1–21.

Heaney-Kieras, J. & D. J. Chapman (1976). Structural studies on the extracellular polysaccharide of the red alga *Porphyridium cruentum. Carbohydr. Res.* 52, 169–177.

Heaney-Kieras, J., L. Roden & D. J. Chapman (1977). Covalent linkage of protein to carbohydrate in the extracellular protein-polysaccharide from the red alga *Porphyridium cruentum. Biochem. J.* 164, 9–17.

Medcalf, D. G., J. R. Scott, J. H. Brannon, G. A. Henerick, R. L. Cunningham, J. H. Chessen, & J. Shah (1975). Some structural features and viscometric properties of the extracellular polysaccharides from *Porphyridium cruentum. Carbohydr. Res.* 44,87–96.

Percival, E. & R. A. J. Foyle (1979). The extracellular polysaccharide of *Porphyridium cruentum* and *Porphyridium aerugineum. Carbohydr. Res.* 72, 165–176.

Ramus, J. (1972). The production of the extracellular polysaccharide by the unicellular red alga *Porphyridium aerugineum. J. Phycol.* 8, 97–111.

Ramus, J. (1973). Cell surface polysaccharides of the red alga *Porphyridium*. In, F. Loewus, (ed.) *Biogenesis of Plant Cell Wall Polysaccharides*. Academic Press, New York, pp. 333–358.

Ramus, J. (1986). Rhodophyte unicells: Biopolymer, physiology and production. In, W. R. Barclay & R. P. McIntosh (eds). *Algae Biomass Technologies. Nova Hedwigia, Beih.* 83. J. Cramer, Berlin, pp. 52–57.

von Witsch, H., A. Bolze, & U. Hornung (1983). Wachstum und Bildung extrazellularer polysaccharide in Batch-Kulturen von *Porphyridium aerugineum*, Rhodophyceae. *Ber. deutsch. bot. Gesells.* 96, 469–481.

NOTES FOR INSTRUCTORS
Supplies for each pair of students

1. 10 mL beakers (12)
2. fine-porosity sinter glass filter, 15 mL (1)
3. 50 mL side-arm filtration flask (1)
4. pkg 20 μL capillary pipettes for chromatographic use (1)
5. Prelaid silica gel G plates – Impregnated* (4)
6. Prelaid silica gel G plates – Unimpregnated** (4)

---

* To impregnate plates: place plate edge down in a chromatographic tank containing 100 mL 0.02 M sodium acetate. Allow to develop to top (about 90 min), remove, dry by heating for 1 h at 100°C. This entire operation should be performed prior to class time so the plates are ready for use.

**The unimpregnated silica gel plate should be activated by heating for 1 h at 100°C. Activate prior to class time, so that the activated plate is ready for use.

7. Regular sized chromatographic tanks (2)
8. Whatman #1 filter paper – sheets
9. Aerosol or air-atomizer chromatographic reagent sprayers (3)
10. 5 mL, 10 mL serological pipettes (6)
11. Vacuum digestion (hydrolysis) tubes (3)
12. Weighing bottle (1)
13. Spatulas (3)

Chemicals

1. $BaCO_3$
2. Methanol
3. Chloroform
4. Ethyl acetate
5. Acetic acid-glacial
6. n-butanol
7. Diethyl ether
8. Cetylpyridinium chloride

Instrumentation

1. Preparative centrifuge with appropriate number of 15 mL, 40 mL tubes and 250 mL bottles
2. Analytical balance
3. Oven to heat chromatographic plates
4. Oven for digestion
5. Air blower or compressed air

Reagent solutions

1. 0.01 M sodium acetate
2. 1 N H_2SO_4
3. 0.1% Xylose solution
4. 0.1% Galactose solution
5. 0.1% Glucose solution
6. 0.1% Glucuronic acid solution
7. 0.2% naphthoresorcinol + 10% phosphoric acid in n-butanol (500 mL)
8. 25 mL *p*-anisaldehyde, 25 mL conc. H_2SO_4, 2.5 mL acetic acid in 450 mL ethanol
9. 0.1 M *p*-anisidine + 0.1 M phthalic acid in 96% ethanol (500 mL)

Cetylpyridinium chloride, sugars and organic reagent solution chemicals obtainable from Sigma Chemical Co. St. Louis, Mo.

Cultures

Three weeks before the experiment, inoculate 1 L of medium with 100 mL of an actively growing culture of *Porphyridium cruentum* (UTEX 161) (or *Porphyridium aerugineum* [UTEX 755]). Grow the cultures in the appropriate medium (see following), with shaking or aeration, at ca. 75 μE m^{-2} sec^{-1} at 18–20° C. Production of polysaccharide can be easily monitored by observing the increase in viscosity. Old cultures (e.g., 6 weeks) will have produced copious polysaccharide, but are so viscous as to be difficult to handle.

P. cruentum *medium*

NaCl	27 g
$MgSO_4 \cdot 7H_2O$	6.6 g
$MgCl_2 \cdot 6H_2O$	5.6 g
$CaCl_2 \cdot 2H_2O$	1.5 g
KNO_3	1.0 g
KH_2PO_4	0.07 g
$NaHCO_3$	40 mg
1 *M* Tris-HCl	20 mL
Fe-solution*	2 mL
Trace elements**	1 mL
H_2O to 1000 mL	

Fe-solution

2.4 g $FeCl_3.6H_2O$ + 1 g Na_2EDTA in 500 mL H_2O

**Trace elements*

$ZnCl_2$	1.2 g
H_3BO_3	1.8 g
$MnCl_2 \cdot 4H_2O$	1.2 g
$CoCl_2 \cdot 4H_2O$	150 mg
Ammonium paramolybdate	370 mg
$CuSO_4 \cdot 5H_2O$	80 mg
H_2O to 1000 mL	

P. aerugineum *medium*

$NaNO_3$	450 mg
KCl	30 mg
$CaCl_2 \cdot 2H_2O$	36 mg
$MgSO_4 \cdot 7H_2O$	100 mg

Na$_2$ glycerophosphate .5H$_2$O	90 mg
Tricine buffer	986 mg
Fe-solution*	1 mL
Trace elements**	10 mL
H$_2$O to 1000 mL	
Adjust to pH 7.6	

Fe-solution

As for *P. cruentum*

**Trace Elements*

Na$_2$EDTA	1 g
H$_3$BO$_3$	200 mg
FeCl$_3$·6H$_2$O	50 mg
MnCl$_2$·4H$_2$O	50 mg
ZnCl$_2$	10 mg
CoCl$_2$·6H$_2$O	5 mg
CuSO$_4$·5H$_2$O	10 mg
H$_2$O to 1000 mL	

Comments

It is suggested that the students work in pairs. Each pair should use only one chromatographic system and one of the reagent sprays. Use as many combinations as possible in the class.

This experiment extends beyond one laboratory session. The instructor can simplify the experiment by replacing the *Porphyridium* polysaccharide with either commercial agar or carrageenan, or a crude sample of carrageenan from the following experiment. If this procedure is followed, omit Steps 1-5 and start at Step 6 with 20 mg agar or 20 mg carrageenan (do not use alginates).

It is recommended that Steps 6–10 be carried out during one lab session, and Steps 11–18 be carried out in a subsequent session. The particular chromatographic systems used may take up to 90 minutes to develop. The neutralised digest (Step 10) can be stored in the refrigerator.

The oven for heating the plates should be located in a fume hood.

The sealed tubes for digestion should be placed in a shielded container when being heated.

Two pairs of students could utilize the same hydrolysate. For each two pairs approximately 1 L of culture should provide sufficient polysaccharide, 10–25 mg, depending upon age.

17

Crude extraction and testing of carrageenan

Herbert Vandermeulen

Huntsman Marine Laboratory, St. Andrew's, NB, Canada
EOG 2X0

FOR THE STUDENT

Introduction

The cell walls of the Rhodophyta are constructed of two primary components, one fibrillar and the other mucilaginous. The mucilaginous component frequently contains polysaccharides, which are of commercial value. These polysaccharides are composed of repeating sequences of galactose derivatives. Their commercial value lies in their ability to act as gelling or suspending agents in solution. These macromolecules produce the "gel" state by formation of coils or spirals as a tertiary structure. Mackie and Preston (1974), Chapman and Chapman (1980) and Abbott and Cheney (1982) have provided reviews of the structure, uses and value of these chemicals.

The purpose of this exercise is to extract the sulfated galactan polysaccharide, carrageenan, from a red alga. Carrageenan occurs in a variety of forms, each having slightly different chemical properties. You will test your carrageenan extract for its ability to suspend cocoa powder in solution.

Procedure

Extraction. Take a small handful of the frozen seaweed, place it into a 1 L beaker of tap water. Clean off the plants with the water, then remove and blot dry on a paper towel. Weigh out 5–10 g of the clean seaweed for extraction. Place the seaweed into another clean, dry 1 L beaker, add 7 mL of 2% NaOH and 343 mL distilled water. Place the beaker onto a hot plate and boil the slurry for 15 min. Place a large watchglass on top to prevent water loss. Stir with a glass rod if the extract starts to foam out of the beaker (use gloves). Watch carefully, because the foaming increases as time goes by.

After 15 min pour the contents of the beaker into a blender (use gloves). Apply a short burst of power to chop the material into small pieces (about 4 mm diameter). *Do not over blend*, as it will turn into a green liquid that is difficult to filter. Pour the hot material back into the beaker. Wash the blender with 150 mL very hot distilled water and add this to the beaker. Simmer the mixture for another 15 min, stirring to keep the froth down. Prepare the filter as outlined below while the solution is boiling.

Set up a 1 L beaker with four layers of cheesecloth taped over its mouth. Pour the hot slurry through the cloth. Stop when you have about 300 mL of filtrate. Before the filtrate cools pour about 100 mL into a large fluted filter paper in a glass funnel. Collect 20 mL of the final product into each of two 100 mL beakers. This should not take longer than 15 min each. The final filtrate should be almost clear. Cool the beakers on ice to about room temperature if the final filtrate is still warm. Discard the remaining material in the funnel. Save the remaining cheesecloth filtrate for carrageenan precipitation.

Suspension properties. Into another 100 mL beaker add 20 mL of distilled water. Add 10 mg of pure cocoa powder to each of the distilled water and one clear algal extract. Stir each solution until all lumps are gone and the cocoa is well mixed. Don't cross contaminate with the stirring rod, wash it after use. Fill a spectrophotometer tube (labeled) with the water/cocoa mixture and record the time. Pour the extract/cocoa mixture into another tube and record the time. Let both tubes sit for exactly 20 min. Carry out the carrageenan precipitation while you wait.

Prepare a blank of distilled water in a spectrophotometer tube. Fill a fourth tube with some of the remaining clear filtrate as an algal extract blank. Record the absorbance of all four tubes at 550 nm. Standardize the results by subtracting each "blank" reading from the appropriate "cocoa" tube.

Carrageenan precipitation. Place 50 mL of the colored cheesecloth filtrate into a 400 mL beaker. Cool the beaker on ice if the filtrate is still warm.

Add 100 mL of *ice cold* isopropyl alcohol while stirring with a glass rod. The precipitate should stick to the glass rod. Discard the precipitate after examining it for texture.

Questions

Which tubes contained the darker solution? What does this tell you about the suspending ability of the extract?

REFERENCES

Abbott, I. A. & D. P. Cheney (1982). Commercial uses of algal products: introduction and bibliography. In, J. R. Rosowski & B. C. Parker (eds.) *Selected Papers in Phycology II.*, Phycological Society of America, Inc., Lawrence, Kansas, pp. 779–787.

Chapman, V. J. & D. J. Chapman (1980). *Seaweeds and Their Uses.* Chapman and Hall, London, 334 pp.

Mackie, W. & R. D. Preston (1974). Cell wall and intercellular region polysaccharides. In, W. D. P. Stewart (ed.) *Algal Physiology and Biochemistry.* University of California Press, Los Angeles, pp. 40–85.

NOTES FOR INSTRUCTORS
Materials for 12 students or groups

1. spectrophotometry test tubes (48)
2. 100 mL beakers (36)
3. one liter beakers (36)
4. weighing boats (36)
5. 400 mL beakers (12)
6. Large watchglasses (17 cm dia.) (12)
7. hotplates (12)
8. glass rods (30 cm long) (12)
9. pairs oven mitts (12)
10. glass funnels (12)
11. 250 mL graduated cylinders (12)
12. 100 mL graduated cylinders (12)
13. 20 mL graduated cylinders (12)
14. grease pencils (12)
15. test tube racks (12)
16. stands with clamp and ring to hold funnel (12)
17. weighing balances, capable of weighing from 10 mg $\pm$ 1 mg to 20 g $\pm$ 1 g (3)

18. blenders (3)
19. spectrophotometers (2)
20. bolt cheesecloth (1)
21. roll masking tape (1) (or 12 large rubber bands)
22. box Whatman #1 filter paper (18.5 cm dia.) (1)
23. freezer to store seaweed (1)
24. 2% NaOH solution by weight (100 mL)
25. pure cocoa powder *(not instant)* (10 g)
26. isopropyl alcohol (1.5 L)
27. ice and container (6 kg)
28. frozen seaweed (0.5–1.0 kg) (if storage time is to be more than a few weeks a −70°C freezer is recommended). Carrageenophytes include *Chondrus, Eucheuma, Gigartina, Gymnogongrus, Hypnea, Iridaea, Pachymenia,* and *Rhodoglossum.*

Scheduling

It is recommended that the experiment be conducted with the students working singly or in pairs so adequate replicates can be produced for quantification. Three hours is sufficient.

Comments

5.0 g wet weight of plant material is recommended for *Iridaea cordata* and *Gigartina exasperata*. More seaweed may be required to produce a visible amount of precipitate if using other species. All students should attempt to produce the precipitate, as this procedure is not always successful and it gives some estimate of the quality of extract produced by the student.

18

Extraction of alginic acid from a brown seaweed

J. N. C. Whyte

Fisheries and Oceans, Pacific Biological Station, Nanaimo, BC, Canada, V9R 5K6

FOR THE STUDENT

Introduction

Salts of alginic acid (alginates) occur in all brown seaweeds as structural components of the cell walls (Percival & McDowell 1967). Commercially extracted alginates, principally from *Macrocystis pyrifera, Ascophyllum nodosum* and species of *Laminaria,* are used for their thickening, stabilizing, film-forming and gel-producing properties (Chapman and Chapman 1980).

The extraction process is based on the conversion of mixed insoluble salts of alginic acid in the cell walls into a soluble salt that is suitable for aqueous extraction. Processing involves leaching the kelp with hot aqueous calcium chloride or cold dilute acid to remove soluble polysaccharides, such as fucoidan (a mix of fucans) and laminaran, polyvalent cations, and low-molecular-weight components, such as mannitol, sugars, amino acids, and peptides, prior to sodium carbonate extraction to yield alginic acid (Finch et al. 1986).

Alginic acid consists of three types of block polymers essentially of β-D-$(1 \rightarrow 4)$-linked mannuronic acid residues, α-L-$(1 \rightarrow 4)$-linked guluronic acid residues and an alternating sequence of these residues as illustrated in Fig. 18.1. The proportion of these blocks is dependent on the algal source of the alginate. Partial acid hydrolysis separates mixed blocks from homopolymer blocks and then the latter can be separated from each other

Alginic Acid

Residues

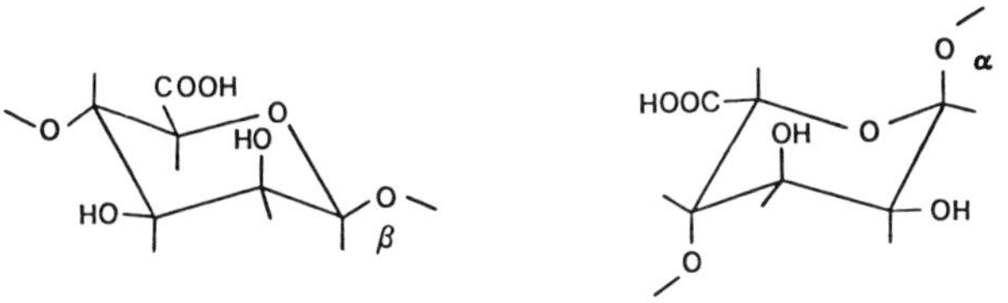

β-D-mannuronic acid   α-L-guluronic acid

Block sequences

M- M- M - M- M- M- M- M- M- M

G- G- G - G- G- G- G- G- G- G

M- G- M- G- M- G- M- G- M- G

Fig. 18.1. Structure of mannuronic acid, guluronic acid, and polymeric blocks of these residues in alginic acid.

by their differential acid solubilities (Haug et al. 1974; Cheshire and Hallam 1985). Properties and functionality of the alginates isolated from different brown algae reflect the conformational structures adopted by these block polymers in solution (Grant et al. 1973).

The aim of this experiment is to extract alginic acid from a brown seaweed, by a process utilizing and illustrating the solubilities of alginic acid and its salts. A flow diagram of the extraction process is presented in Fig. 18.2

Procedure

Removal of soluble and storage polysaccharides. Add seaweed (100 g blended fresh or 10 g ground dry) and aqueous calcium chloride (1%, 300 mL) to a 1 L beaker on a hotplate. Stir the contents using the high-torque electric stirrer and heat at 60°C for 15 min. Pour the mixture into two centrifuge bottles, using mitts, position in centrifuge cups and balance, then centrifuge at 2000 rpm for 5 min. Discard the supernatant and return the seaweed residue to the 1 L beaker for reextraction with a further portion of hot aqueous calcium chloride. Combine the residues from the two

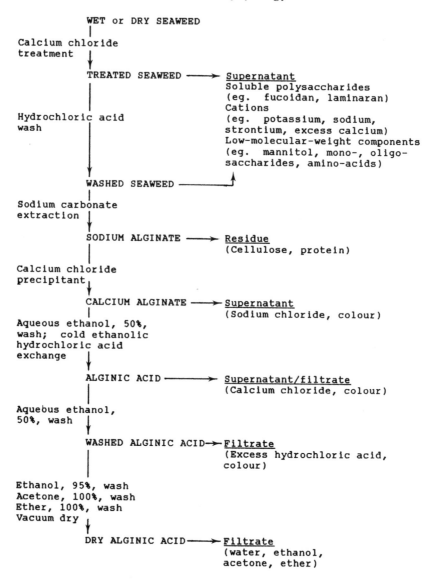

Fig. 18.2. Flow diagram of alginic acid extraction process.

centrifuge bottles into one bottle to subsequently wash with water and
dilute acid. Add enough water to fill the centrifuge bottle up to the ¾
point, mix contents thoroughly using a spatula, and centrifuge for 5 min.
Discard the supernatant, add hydrochloric acid (0.05 *M*) to fill the bottle
up to the ¾ mark, mix the contents and then centrifuge. Wash residual

seaweed twice more with hydrochloric acid, discarding supernatants each time.

Extraction of alginates. Return the residue to the 1 L beaker using aqueous sodium carbonate (0.5%) to effect total transfer. Add the aqueous sodium carbonate (0.5%) for a total of 300 mL, stir contents with electric stirrer, heat at 45°C for 30 min., and then centrifuge. Save the supernatant. Meanwhile, weigh the glass-fibre filter (GF/A) and clamp in Millipore filter holder. Return residue to the 1 L beaker and digest with a further 300 mL aqueous sodium carbonate at 45°C for 15 min., then centrifuge. Dilute combined centrifugates with distilled water to 1 L total volume; centrifuge this sodium alginate solution for clarification if necessary. Pour alginate solution slowly with stirring (use glass rod) into the aqueous calcium chloride (25%, 100 mL) contained in a 2 L beaker and continue stirring for an additional 10 min. Centrifuge precipitated calcium alginate to aggregate and discard supernatant. Transfer the calcium alginate to preweighed filter in filter holder and filter under vacuum. Wash the calcium alginate first with 50% aqueous ethanol (200 mL), then slowly with cold 1 M hydrochloric acid, as a 50% aqueous ethanolic solution (500 mL). To improve the rate of exchange, gently stir the fibrous polysaccharide and press it against the side of the filter holder using the glass rod. Wash the product, alginic acid, (still on the filter paper) with 50% aqueous ethanol until free from chloride; test the filtrate with one drop saturated silver nitrate solution until no formation of chloride precipitate is evident. Wash with 95% ethanol (200 mL), acetone (200 mL), and diethyl ether (100 mL) for solvent exchange, and dry *in vacuo* in a desiccator for 15 min. Weigh on filter and calculate precentage yield of partially purified alginic acid.

Assessment of purity. Weigh alginic acid (200 mg) into a 100 mL beaker, pipette in 25 mL standardized aqueous sodium hydroxide (0.1 M), and stir using the magnetic stirring bar on the hotplate stirrer at 30°C until total dissolution of acid. Back-titrate excess alkali with standardized hydrochloric acid (0.1 M) using phenolphthalein as the indicator (red to colorless). Calculate the percentage purity of the alginic acid given that 1 mL 0.1 M sodium hydroxide is equivalent to 0.0194 g alginic acid in the hydrated form in which it exists even after solvent exchange.

REFERENCES

Chapman, V. J. & D. J. Chapman (1980). *Seaweeds and Their Uses.* Chapman and Hall, London, 334 pp.

Cheshire, A. C. & N. D. Hallam (1985). The environmental role of alginates in *Durvillaea potatorum*, Fucales, Phaeophyta. *Phycologia* 24, 147–154.

Finch, P., E. Percival, I. R. Slaiding & H. Weigel (1986). Carbohydrates of the Antarctic brown seaweed *Ascoseira mirabilis*. *Phytochemistry* 25, 443–448.

Grant, G. T., E. R. Morris, D. A. Rees, P. J. C. Smith & D. Thom (1973). Biological interactions between polysaccharides and divalent cations: the egg-box model. *FEBS Letters* 32, 195–198.

Haug, A., B. Larsen & O. Smidsrød, (1974). Uronic acid sequence in alginate from different sources. *Carbohyd. Res.* 32, 217–225.

Percival, E., & R. H. McDowell (1967). *Chemistry and Enzymology of Marine Algal Polysaccharides*. Academic Press, New York, 219 pp.

NOTES FOR INSTRUCTORS

Materials and equipment required (24 students)

1. 1 L heavy duty beakers (safer when stirring hot mixtures) (12)
2. 2 L beakers (12)
3. hot plate stirrers (12)
4. thermometers, 0°–100°C (12)
5. plastic weigh boats, standard size (12)
6. balance (50 g–0.0001 g) (1)
7. high-torque electric stirrers fitted with stainless steel stirrer shafts preferably with Teflon-coated blades (12)
8. centrifuges, horizontal heads, four cups with 250 mL centrifuge bottles (two or four pairs can centrifuge at once as space dictates) (3)
9. glass rods (0.8 × 25 cm) (12)
10. oven mitts (12)
11. Millipore filter holders with box of 5.5 cm GF/A filters (12)
12. desiccators fitted with Tygon pressure tubing to fit aspirator or Nalgene polypropylene water pump (6)
13. 25 mL pipettes (12)
14. 25 mL burettes (12)
15. heavy teflon-coated magnetic stirring bar (spinbar type) (12)
16. trip balance, single beam (for balancing centrifuge cups containing bottles prior to centrifugation) (1)
17. brown algae (species as mentioned in Introduction), 150 g dry powdered (using a mill or mortar and pestle), or 1500 g wet (blended in Waring blendor)

The following volumes are required (measurements are in slight excess of class requirements):

1. 1% aqueous calcium chloride (7.5 L)
2. 0.05 M hydrochloric acid (7.5 L)
3. 0.5% aqueous sodium carbonate (7.5 L)
4. 25% aqueous calcium chloride filtered (2 L)
5. 80% aqueous ethanol (15 L)
6. 1.0 M hydrochloric acid as a 50% aqueous ethanolic solution (cool in ice-water bath during experiment) (6 L)
7. phenolphthalein indicator (0.5% in 50% aqueous ethanol) (10 mL)
8. saturated aqueous silver nitrate (5 mL)
9. 95% ethanol (3 L)
10. 100% acetone (3 L)
11. 100% diethyl ether (1.5 L)

Recommendations for scheduling

This experiment should be conducted with students in pairs. A period in excess of three hours will be required, particularly if access to the centrifuge is restricted. Stages in the extraction process most suitable for carry-over to the next period are when calcium alginate, or washed alginic acid, or dry alginic acid is isolated (see flow diagram, Fig. 18.2). Leaving alginates in acidic or basic solutions should be avoided, as depolymerization will occur.

Comments

Ethanol, acetone and ether should be used in a fume hood. Opened, partially used cans of ether are prone to explode (see precautions in Exp. 9 procedures).
Phenolics will condense onto the alginates, giving a reddish-brown color in alkaline solution which largely disappears in acidic solution.

Life history phases in natural populations of Gigartinaceae (Rhodophyta): quantification using resorcinol

David J. Garbary

Department of Biology, St. Francis Xavier University,
Antigonish, Nova Scotia, Canada B2G 1C0

Robert E. DeWreede

Department of Botany, University of British Columbia,
Vancouver, BC, Canada V6T 2B1

FOR THE STUDENT

Introduction

In autecological studies it is often important to have information on the relative abundance of the various life history phases in a population or group of populations. This is often difficult to quantify in plants with isomorphic gametophytic and tetrasporophytic generations because of the presence of many nonreproductive plants. In Gigartinaceae, it is possible to characterize the life history phase of a plant based on the fractionation of carrageenans since kappa-carrageenan is produced by gametophytes and lambda-carrageenan by tetrasporophytes (McCandless et al. 1973). Craigie and Leigh (1978) outlined a chemical method using the reagent resorcinol to assess, quantitatively and qualitatively, the carrageenans in a single plant so that its life history phase could be identified. This method

174

was modified by Dyck et al. (1985) to produce a simpler method for the qualitative determination of lambda- or kappa-carrageenan. Their method allows an investigator to study the temporal and spatial abundance of gametophytic and tetrasporophytic phases in Gigartinaceae. The method is particularly useful in that it does not require destructive sampling of the population; only small portions of each blade are needed.

Procedure

(to be carried out in a fume hood)

1. Place a single sample disk (either air dried or after blotting with paper towel) in a test tube.
2. Add 2 mL of resorcinol-acetal reagent. DO NOT PIPETTE BY MOUTH!
3. Place in water bath for 60 sec. at 80°–90°C.
4. Remove from water bath and note color of reagent mixture. Lack of color or very slight reaction indicates blade was tetrasporophytic; pink to red reactions indicate blade was gametophytic.
5. Record data.
6. If necessary, dispose of sample/reagent mixture and wash tube before proceeding to next sample.

Questions and exercises

1. Assuming equal survival of tetraspores and carpospores, and equivalent spore output by cystocarpic and tetrasporic blades, what proportions of gametophytic and tetrasporophytic thalli might be expected in populations in equilibrium? Note: in dioecious species, four spores from a single tetrasporangium produce two male and two female plants.
2. Repeat preceding problem but assume that tetrasporic blades produce twice as many spores as gametophytic blades. How would this change in monoecious species?
3. The preceding exercises are oversimplifications of processes that would control or limit the relative abundance of gametophytic and sporophytic blades in a population. What other factors or processes might be involved?
4. Design a research program to investigate the dynamic aspects of life history phases of a member of the Gigartinaceae in nature. How would the research program differ in a species where the resorcinol test could not be used?

REFERENCES AND SUGGESTED READING

Craigie, J. S. & J. D. Pringle (1978). Spatial distribution of tetrasporophytes and gametophytes in four Maritime populations of *Chondrus crispus. Can. J. Bot.,* 56, 2910–2914.

Craigie, J. S. & C. Leigh (1978). Carrageenans and agars. In, J. A. Hellebust & J. S. Craigie (eds) *Handbook of Phycological Methods: Physiological and Biochemical Methods.* Cambridge University Press, London, pp. 109–131.

Dyck, L., R. E. DeWreede & D. Garbary (1985). Life history phases in *Iridaea cordata* (Gigartinaceae): relative abundance and distribution from British Columbia to California. *Jap. J. Phycol.* 33, 225–232.

Hansen, J. E. (1977). Ecology and natural history of *Iridaea cordata* (Gigartinales, Rhodophyta): growth. *J. Phycol.* 13, 285–294.

Hansen, J. E. & W. T. Doyle (1976). Ecology and natural history of *Iridaea cordata* (Rhodophyta, Gigartinaceae): populations structure. *J. Phycol.* 12, 273–278.

Mathieson, A. C. & R. L. Burns (1975). Ecological studies of economic red algae. V. Growth and reproduction of natural and harvested populations of *Chondrus crispus* Stackhouse in New Hampshire. *J. Exp. Mar. Biol. Ecol.* 17, 137–156.

McCandless, E. L., J. S. Craigie & J. A. Walter (1973). Carrageenans in the gametophytic and sporophytic stages of *Chondrus crispus. Planta* 112, 201–212.

NOTES FOR INSTRUCTORS

Materials

Species. Two species ideally suited to this exercise are *Chondrus crispus* Stackhouse (in the Atlantic) and *Iridaea cordata* (Turner) Bory (in the Pacific), both members of the Gigartinaceae. Petrocelidaceae (e.g., *Mastocarpus*) show a similar fractionation of carrageenan types, but in heteromorphic life histories. *Eucheuma* spp. (Solieriaceae) are not suitable because they do not have different carrageenan types in gametophytic and tetrasporophytic phases.

Sampling procedure. Sampling can be done well in advance of the laboratory by the instructor or by the students who will analyze the samples at a later date. Sampling of plants can be carried out using a single-hole paper punch, a cork borer, or a leather punch with a rotating head of different size punches (use the largest one). This provides a uniform sample disk for the analysis (ca. 3 mg dry weight for *Iridaea cordata*). The instructor (or student) designates a population to be sampled and works through the population collecting the disks. Sampling design and subsequent statistical analyses should be specified by the instructor. A transect method may be preferred if there is concern about resampling the same individual (i.e., a single plant with multiple upright blades). A minimum of 50 disks per population should be used.

The algal disks can be processed immediately (after blotting to remove excess moisture), or air dried and stored in vials for future analysis. When the collecting is carried out, be sure to collect a sample of disks (ca. 50 disks) from both cystocarpic and tetrasporic blades. In each batch of samples analysed, one of each of these 'known' disks should be tested, thus providing a control or standard color reaction for each batch.

Supplies

1. Single hole paper punch (one per student group if they are doing the collection of disks). Note: cork borers (be sure to have a surface against which the disks may be cut) or leather punches (5–7 mm diameter) may also be used
2. paper towels for drying the disks
3. vials for storing disks after drying (one per population sampled)
4. test tubes, 10 mL (10–25 per group of students)
5. test tube racks (one per student group)
6. water bath (at 80°–90°C, one per student group)
7. hydrochloric acid
8. resorcinol (see following)
9. acetal (1, 1 - dioxyethane)
10. pipettes and bulbs (minimum of two per student group) (an automatic dispenser is highly desirable) – Pasteur pipettes can be marked for 1 mL

Recipe for test reagent

1. Resorcinol stock solution (keeps one week)
   a. Add 150 mg resorcinol to 100 mL distilled water
   b. Add 9 mL of stock solution (1a) to 100 mL concentrated HCl
2. Acetal stock solution (keeps three weeks)
   a. Add 0.1 mL of acetal to 10 mL distilled water
   b. Dilute 1 mL of 2a in 25 mL distilled water

Add 1 mL of diluted solution (2b) to (1b). This is the resorcinol-acetal test reagent that is added to the disks. It should be made fresh daily. A separate bottle of test reagent should be available for each student group.

Scheduling

For class use, students can be divided into a number of groups such that each group is given a separate population or sample to analyze. We have found that having two or three people in a group makes the analysis proceed faster, with separate individuals preparing the tubes, recording the results and washing the tubes. The samples might represent: (1) populations along an environmental gradient (e.g., salinity, wave exposure, substratum), (2) samples from different heights in the intertidal and/or subtidal regions, (3) samples from different parts of a subtidal bed, (4) samples from different months or years, (5) samples from different size classes, or (6) reproductive versus nonreproductive blades.

The analysis and exchange of data among the various groups can be easily carried out within one to two hours. Depending upon the complexity of the experimental design and statistics required for analysis, it is possible within a three-hour laboratory session to have a class discussion of the results prior to preparation of a report. The questions and exercises in the preceding section might provide items for class discussion or provide the basis of material to be included in a report. The *Polysiphonia*-type life history of Gigartinaceae should be understood before starting the analysis. A number of references are included with the laboratory that deal either with the method directly, or with investigations where it has been used or might have been useful. These papers deal with aspects of the life history of *Chondrus* and *Iridaea* in natural populations.

Note: Batches of disks can be processed in lots of 25 to 75 tubes. Larger batch size (and too much plant material in a sample) may result in a slight color reaction in tubes containing tetrasporophytic disks.

Optional

A spectrophotometer can be used to verify the absorption spectra for the reactions given by Dyck et al. (1978). Dilute the reaction mixture with an equal part of the resorcinol reagent before taking readings. It is also possible to quickly check any sample that might have an equivocal color reaction. Take readings at 415 and 500 nm. In gametophytic samples, the absorbance should be much greater at 500 nm. In tetrasporophytes, the absorbance value for 415 nm should be greater than or very close to the value for 500 nm.

20

Sulfate uptake in larger marine algae

L. V. Evans

Department of Plant Sciences, University of Leeds, Leeds LS2 9JT, United Kingdom

FOR THE STUDENT

Introduction

Sulfate is an important macronutrient in algae. It is used in the formation of major sulfated polysaccharides such as fucoidan, one of a family of glucuronoxylofucans in the brown algae, and agar and carrageenan, sulfated galactans of the red algae. A spectacular result of sulfate uptake is the high acidity of the brown macroalga *Desmarestia,* which is due to high concentrations of sulfuric acid in the cell vacuoles. Sulfate is also used for partial reduction to sulfolipid and other sulfonic acids, and for complete reduction to give the thiol groups of the amino acids cysteine and methionine as well as other important molecules such as coenzyme A, thiamine and biotin. Sulfate uptake is the first step in sulfate utilization and is an active process following Michaelis-Menten kinetics, with an apparent K_m for sulfate generally in the range 10^{-5} to 10^{-4} M (Schiff & Hodson 1973; Tsang, Hodson & Schiff 1978), although lower values are reported for some cells (see e.g., Ramus 1974; Millard & Evans, 1982).

Where radiolabeled sulfated compounds are not secreted into the medium in significant amounts by the cells or tissue during the period of the experiment, the simplest method of measuring sulfate uptake in algae is to incubate with an appropriate specific activity of sulfate and to measure disappearance of this from the medium with time. This method works well with microalgae such as *Chlorella,* where relatively short incubation periods (2 h) may be used (see e.g., Hodson, Schiff & Scarsella 1968;

179

Tsang et al., 1976). However, with the longer incubation times needed for macroalgae it is necessary for the investigator to be aware that problems associated with this technique could lead to inaccuracies, particularly if the method is used for determining values for kinetic parameters (V_{max} and K_m). (See Harrison & Druehl (1982) for an overview of these problems.) For example, the tissue is likely to exhaust available N and P over a prolonged period, thus affecting its metabolic state and sulfate uptake. Reduction in dissolved inorganic carbon (DIC) also occurs over a relatively short time (2 h) (see e.g., Bidwell & McLachlan 1985), resulting in reduced photosynthesis which would affect sulfate uptake. There is also the problem of distinguishing between free-space and cellular uptake. The alternative is to measure the amount of radioactive sulfate taken up and accumulated by the cells or tissue. Although this method is preferable for obtaining accurate kinetic data, it would involve the use of more elaborate techniques than are appropriate in the present context. (See Ramus & Groves 1972; Ramus 1974 or Millard & Evans 1982 for the use of this approach with the unicellular red algae *Porphyridium* and *Rhodella* respectively; and Quatrano & Crayton 1973 or Coughlan 1977 for its use with the multicellular brown alga *Fucus*.)

The present experiment is concerned with determining uptake rates by measuring loss of ^{35}S-labeled sulfate from the medium, in different seaweed species or in seaweed tissue of different age, in the presence of varying concentrations of carrier (nonradioactive) sulfate. The concentration of ^{35}S used is very low, because it is carrier-free (every atom radioactive). The rate of uptake is also very low, as it follows Michaelis-Menten kinetics. Increasing the concentration of carrier sulfate results in increased uptake of both ^{32}S and ^{35}S, but the concurrent dilution of radioactive sulfate results in a decreased uptake of ^{35}S (which is being measured). The experiments allow you to determine an optimal concentration of carrier sulfate to give a compromise between rate of uptake and isotope dilution.

Procedure

CAUTION: This experiment involves the use of radioactive sulfate. Your instructor will counsel you on working with radioisotopes.

1. Excise tips of a brown alga such as *Pelvetia* or *Fucus* 1 cm from the apex. Choose epiphyte-free areas from the lower third of the fronds of *Laminaria* species or from the distal ends of thalloid red algae such as *Palmaria* or *Chondrus,* and cut discs (e.g., cork borer No. 6) or 1 cm^2 pieces. A total of five samples of the alga to be used are required; weigh and adjust until each is approximately 1 g.
2. Wash the material extensively in sterile seawater or Instant Ocean, then

in sterile ASP12 medium in which MgSO$_4$ has been replaced with MgCl$_2$. Leave in this overnight if possible.

3. Into four numbered Erlenmeyer flasks put 20 mL sterile sulfate-free ASP12 medium, and into a fifth numbered flask the same volume of sterile seawater or Instant Ocean. Lightly blot the seaweed pieces, reweigh quickly and accurately (this time recording weights) and transfer accurately-weighed, approximately 1 g samples to each of flasks 1–5.

4. Add 5 μCi carrier-free Na$_2$^{35}SO$_4$ to each flask, swirl vigorously and immediately take from each flask three aliquots of 0.2 mL and transfer to numbered scintillation vials for counting (see Step 7). These are the zero time samples.

5. To flasks 2–4 add carrier sulfate (as Na$_2$^{32}SO$_4$) so as to give final concentrations as follows:

Flask 2: $10^{-5}M$ (add 0.1 mL of a 2 mM solution to 20 mL medium)
Flask 3: $10^{-4}M$ (add 0.1 mL of a 20 mM solution to 20 mL medium)
Flask 4: $10^{-3}M$ (add 0.1 mL of a 200 mM solution to 20 mL medium)

Carrier sulfate should not be added to flask 1 (carrier sulfate-free medium) or flask 5 (seawater or Instant Ocean). The SO$_4^{2-}$ content of seawater is approximately 28 mM.

6. Place the flasks on a shaker at approximately 15°C and an irradiance level of 500–1000 μE m^{-2} s^{-1} from cool white fluorescent tubes.

7. At intervals, remove three aliquots of 0.2 mL from each flask and transfer to numbered scintillation vials for counting. (The intervals can be planned so as to fit in with your other commitments. The first sample should be taken at 2 or 3 h after the zero time sample (step 4) and the last after 24-30 h, with at least two in between, e.g. after 6 and 18 h.) To each scintillation vial add 0.3 mL distilled water and 10 mL scintillation fluid. Leave tubes with the instructor for counting in a liquid scintillation spectrometer.

8. Sulfate uptake is calculated by disappearance of radioactivity from the external medium with time; the zero time samples (Step 4) serve to define the initial value for radioactivity in each flask.

9. By subtraction, determine the amount of radioactivity that has been taken up from the medium by the tissue in each of flasks 1–5 at each sampling time. Finally, on a single graph, plot radioactivity incorporated (calculate as cpm min^{-1} g^{-1} fresh weight of material) against time in hours. Use different symbols to represent each flask and record the carrier sulfate concentration used alongside the appropriate symbol.

Questions

1. What is the optimal carrier-sulfate concentration for uptake of radio-labeled sulfate in your system?

2. Why is the uptake rate at 1mM carrier-sulfate very different from that at 100 or 10 μM?
3. Comment on the uptake of radiolabeled sulfate by the material in seawater (or Instant Ocean).
4. How do your results compare with experiments on similar systems reported in the literature (see e.g., Loewus, Wagner, Schiff & Weistrop 1971 and Evans, Simpson & Callow 1973)?
5. What do you think has been the fate of the labeled inorganic sulfate taken up by the alga?
6. How would you proceed to confirm this and to determine how much labeled sulfate has been incorporated in this way?
7. Suggest some practical applications for $^{35}SO_4^{2-}$ uptake in seaweeds (see e.g., Tveter-Gallagher, Cheney & Mathieson 1981).

REFERENCES

Bidwell, R. G. S. & J. McLachlan (1985). Carbon nutrition of seaweeds: photosynthesis, photorespiration and respiration. *J. Exp. Mar. Biol. Ecol.* 86, 15–46.

Coughlan, S. J. (1977). Sulphate uptake in *Fucus serratus. J. Exp. Bot.* 28, 1207–1215.

Evans, L. V., M. E. Callow., E. Percival & V. Fareed (1974). Studies on the synthesis and composition of extra-cellular mucilage in the unicellular red alga *Rhodella. J. Cell Sci.* 16, 1–21.

Evans, L. V., M. Simpson & M. E. Callow (1973). Sulphated polysaccharide synthesis in brown algae. *Planta* 110, 237–252.

Harrison, P. J. & L. D. Druehl (1982). Nutrient uptake and growth in the Laminariales and other macrophytes: a consideration of methods. In, L. M. Srivastava (ed.) *Synthetic and Degradative Processes in Marine Macrophytes.* Walter de Gruyter, Berlin, pp. 99–120.

Hodson, R. C., J. A. Schiff & A. J. Scarsella (1968). Studies of sulfate utilization by algae. 7. *In vivo* metabolism of thiosulfate by *Chlorella. Plant Physiol.* 43, 570–577.

Loewus, F., G. Wagner, J. A. Schiff & J. Weistrop (1971). The incorporation of ^{35}S-labeled sulfate into carrageenan in *Chondrus crispus. Plant Physiol.* 48, 373–375.

Millard, P. & L. V. Evans (1982). Sulphate uptake in the unicellular marine red alga *Rhodella maculata. Arch. Microbiol.* 131, 165–169.

Quatrano, R. S. & M. A. Crayton (1973). Sulfation of fucoidan in *Fucus* embryos. *Dev. Biol.* 30, 29–41.

Ramus, J. (1974). *In vivo* molybdate inhibition of sulfate transfer to *Porphyridium* capsular polysaccharide. *Plant Physiol.* 54, 945–949.

Ramus, J. & S. T. Groves (1972). Incorporation of sulfate into the capsular polysaccharide of the red alga *Porphyridium. J. Cell Biol.* 54, 399–407.

Schiff, J. A. & R. C. Hodson (1973). The metabolism of sulfate. *Annu. Rev. Plant Physiol.* 24, 381–414.

Tsang, M. L-S., R. C. Hodson & J. A. Schiff (1978). Sulfate uptake. In, J. A. Hellebust & J. S. Craigie (eds) *Handbook of Phycological Methods*. Cambridge University Press, pp. 419–425.

Tveter-Gallagher, E., D. Cheney & A. C. Mathieson (1981). Uptake and incorporation of ^{35}S into carrageenan among different strains of *Chondrus crispus*. In, T. Levring (ed.) *Xth Internat. Seaweed Symp.*, Walter de Gruyter & Co, Berlin, pp. 521–530.

NOTES FOR INSTRUCTORS

Requirements for each working group

1. 25 μCi radioactive sulfate, as carrier-free $Na^{35}SO_4$ (half-life 87 days)
2. carrier sulfate, as $Na_2{}^{32}SO_4$ solutions (10 mL of each) at the following concentrations: 2 mM ($2\times10^{-3}M$), 20 mM ($2\times10^{-2}M$), 200 mM ($2\times10^{-1}M$).
3. suitable selection of fresh brown and/or red seaweeds (see following).
4. dissecting instruments, including razor blade or scalpel, forceps etc.
5. cork borers
6. 250 mL conical flasks (for washing material) (2)
7. 500 mL sterile seawater or Instant Ocean
8. 500 mL ASP12 medium without sulfate (see recipe, which follows)
9. 50 mL Erlenmeyer flasks (5)
10. pipettes for dispensing the following volumes: 20 (or 10) mL (1), 0.3 mL (1), 0.2 mL (5), 0.1 mL (4). (One preset automatic pipette with changeable tips could be used for each of these volumes, if available.)
11. automatic dispenser to discharge 10 mL aliquots of scintillation fluid
12. scintillation vials (75) (would give 5 sample times in triplicate from five experimental flasks) and marker pen to number these
13. distilled water
14. 1 L scintillation cocktail: 8 g L^{-1} 2,5-diphenyloxazole (PPO) in toluene: Triton X-100 (2:1). All reagents should be of scintillation grade.

For the class

1. good analytical balances
2. liquid scintillation spectrometer
3. automatic pipette for dispensing radiolabeled sulfate
4. temperature-controlled shaker(s) (preferably gyratory) to run at 15°C
5. light source (e.g., cool white fluorescent tubes) capable of giving an irradiance level of 500–1000 μE m^{-2}s^{-1}

Recipe for sulfate-free ASP12 medium

(Quantities L^{-1} distilled water)

NaCl	28.0 g
KCl	0.7 g
$MgCl_2 \cdot 6H_2O$	4.0 g
$CaCl_2$	400 mg
$NaNO_3$	100 mg
Na_2glycerophosphate $\cdot 5H_2O$	10 mg
PII Metals*	10 mL
SII Metals**	10 mL
K_3PO_4†	1 mL
$Na_2SiO_3 \cdot 5H_2O$†	10 mL
B_{12}/Biotin†	10 μL
Thiamine†	1 mL
Nitriloacetic acid (not essential)	100 mg
Tris (hydroxymethyl)-methylamine (add last)	1.0 g

* PII Metals (to make 100 mL)

$Na_2EDTA \cdot H_2O$	128 mg
$FeCl_3 \cdot 6H_2O$	4.8 mg
H_3BO_3	114 mg
$MnCl_2 \cdot 4H_2O$	14.4 mg
$ZnCl_2$	1.04 mg
$CoCl_2 \cdot 6H_2O$	4.0 mg

** SII Metals (to make 100 mL)

NaBr	1.287 g
$SrCl_2 \cdot 6H_2O$	0.609 g
$LiCl \cdot H_2O$	0.174 g
$NaMoO_4 \cdot 2H_2O$	0.126 g
KI	1.3 mg
RuBr	0.387 g

† The following should be made up separately and added as individual solutions:
K_3PO_4: 1 g in 1 L (10 mg mL^{-1}); add 1 mL L^{-1}
$Na_2SiO_3 \cdot 5H_2O$: 2.6 g in 100 mL (26 mg mL^{-1}); add 10 mL L^{-1}
B_{12}/Biotin: 0.002 g B_{12} + 0.01 g Biotin in 100 mL; add 10 μL L^{-1}
Thiamine: 0.01 g in 100 mL; add 1 mL L^{-1}

Algae

Each member (or pair) from a class can be given a different red or brown seaweed for ^{35}S-labeled sulfate uptake study. It is also interesting to replicate, with two or three groups doing the same tissue, from a range being studied. Tissue undergoing rapid vegetative growth or reproductive tissue can also be compared with mature (non-growing) tissue, or sporophytic material can be compared with gametophytic material, or even the same species from different sources may be compared. (For details of a similar experiment involving the use of a unicellular green alga [*Chlorella pyrenoidosa*], see Hodson et al. [1968] or Tsang et al. [1978]; and for a unicellular red alga [*Rhodella maculata*] see Evans, Callow, Percival & Fareed [1974].)

Scheduling

The experiment cannot be completed in a typical 3 h lab. Only zero-time and 2–3 h samples can be taken in this period. It is therefore necessary for the instructor to make arrangements for students to have access to the lab to take further samples at times convenient to them, as suggested in Step 7.

21

Determining phosphate uptake rates of phytoplankton

P. J. Harrison

Dept. of Oceanography, University of British Columbia,
Vancouver, BC V6T 2B1, Canada

FOR THE STUDENT

Introduction

The rate at which phytoplankton take up phosphate depends on their nutritional history. When phytoplankton are growing under nutrient saturated conditions, the uptake rate is equal to the growth rate when both are expressed in units of time^{-1}. But as cells become nutrient-starved or -limited, the potential uptake rate and growth rate become uncoupled. When these nutrient-limited cells are exposed to an addition of the limiting nutrient, their maximal nutrient uptake rate can be up to several orders of magnitude higher than the maximal growth rate. This enables them to overcome their nutrient debt quickly. The magnitude of this enhanced maximal uptake rate depends on the nutrient in question, the time period over which it is measured, etc. (see Harrison et al., in press, for review).

Uptake rates can be determined directly by using isotopes such as ^{32}P or indirectly by measuring the disappearance of the nutrient from the medium. The former approach is better for field conditions where cell densities are low, while the latter is easier in the laboratory where cell densities can be adjusted so they are $>10^8$ cells L^{-1}. If short incubation times are used, rate measurements using an isotope represent gross uptake rate, while the disappearance method measures net uptake rate (influx minus efflux).

186

Uptake rates must be normalized to a biomass parameter. Uptake rates per cell (μmol 10^6 cells^{-1} h^{-1}) are frequently referred to as transport rates (ρ), while rates that are normalized to the cellular content of the limiting nutrient are referred to as specific uptake rates, V, (h^{-1}). The specific rate is derived by dividing μmoles of nutrient taken up per hour by cell quota, Q, (amount of limiting nutrient per cell). This relationship can be expressed by the following equation: $\rho/Q = V$. In the freshwater literature, the μmoles of nutrient taken up per hour is often divided by the external phosphate concentration in the water, giving a term called "turnover time." Transport rates are usually used in comparing competitive advantages among species, while specific uptake rates are used as a physiological index for species.

Nutrient ions generally enter plant cells by three mechanisms: passive diffusion, facilitated diffusion, and active transport (Fig. 21.1). If transport occurs solely by passive diffusion, the transport rate is directly proportional to the external concentration (Fig. 21.1a). In contrast, facilitated diffusion and active transport exhibit a saturation of the membrane carriers as the external concentration of the ion increases. The relationship between uptake rate of the ion and its external concentration is generally described by a rectangular hyperbola, similar to the Michaelis-Menten equation for enzyme kinetics (Fig. 21.1b):

$$V = V_{max}S/(K_s + S)$$

K_s is the half-saturation constant and it is the substrate concentration at which the uptake rate is half its maximum. The lower the value of K_s, the higher the affinity of the carrier site for the particular ion.

The transport capabilities of a particular phytoplankter are generally described by the parameters V_{max} and K_s. It is important to realize that these parameters do not have a constant value for a species. They vary as a result of nutritional past history, light, temperature and other nutrient interactions. Healey (1980) suggested that the slope of the initial part of the hyperbola (the linear portion) may be a more useful parameter for comparing the competitive ability of various species for a limiting nutrient; this is similar in concept to the use of the slope, α, in the P vs. I curve (Exp. 8). One of the problems in using K_s is that its value is not independent of V_{max} (i.e., when V_{max} decreases, the value of K_s will also decrease even though the initial slope of the hyperbola remains the same) (Fig. 21.1b). The kinetic parameters may be estimated graphically from the rectangular hyperbola, but generally the Michaelis-Menten equation is rearranged to yield a straight line from which K_s and V_{max} can be calcu-

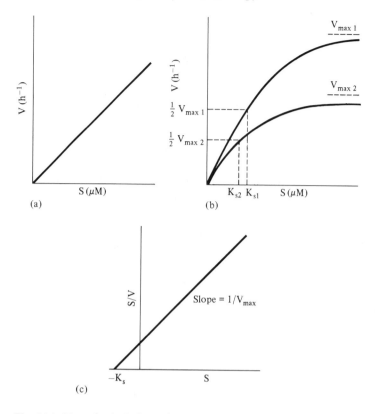

Fig. 21.1. Hypothetical plots of nutrient uptake rate (V) and concentration of the limiting nutrient (S), for (a) passive diffusion only, where V is directly proportional to S; (b) facilitated diffusion or active transport in which V_{max} in example 2 is half V_{max} of example 1, resulting in a concomitant decrease in the K_s; (c) linearized plot of the data in (b) to illustrate how the kinetic parameters V_{max} and K_s are determined graphically. (from Lobban *et al.* 1985, *The Physiological Ecology of Seaweeds,* Cambridge University Press)

lated more accurately by linear regression (Fig. 21.1c). The advantages and disadvantages of the three possible linear plots have been examined by Dowd & Riggs (1965); the S/V versus S or the V/S versus V plot are generally superior to the 1/S versus 1/V plot. Details of uptake methods and data analyses can be found in Healey (1978) and Harrison and Druehl (1982).

The objective of this experiment is to determine phosphate uptake rates for two cultures with different nutritional histories (nutrient-saturated or nutrient-starved).

Procedure

1. Experimental set-up. The instructor has prepared two cultures; one is phosphate-saturated and the other is phosphate-starved (about 24 to 48 h starvation).

    a) Determine the cell density in each culture (as in Exp. 1). These data will be used to normalize the uptake rates on a per-cell basis.

    b) To determine the initial phosphate concentration in the P-starved culture, take duplicate 25 mL samples and filter off the cells as described below. Add phosphate (0.3 mL of 1 μmol mL^{-1}) to the P-starved culture (100 mL) to give a concentration of 3 μM. Shake the culture vigorously. Incubate under the same temperature and irradiance as the culture was growing previously. (NOTE: make your addition carefully because this will be used as your starting concentration in the medium at T = 0). Record the exact time that you make the addition. Take duplicate 25 mL samples at T = 5 and 15 min later; mix the culture before sampling. Immediately filter these samples through a membrane or glass fiber filter to remove the phytoplankton cells, cover the filtrate and store in the refrigerator until all the samples have been taken. Discard the filter containing the cells.

    c) The P-saturated culture already has sufficient phosphate and therefore no addition is necessary. Take duplicate 25 mL samples at T = 0 and 60 min and process the samples as in (b) above. Mix the culture every 15 min so the cells will not settle to the bottom of the flask.

2. After all the samples have been taken, analyze them for phosphate according to Strickland & Parsons (1972) or Parsons et al. (1984). The method is briefly described as follows. The principle of the method is that a seawater sample is allowed to react with a composite reagent containing molybdic acid, ascorbic acid and trivalent antimony. The resulting complex is reduced to give a blue solution that is measured at 885 nm.

    a) Warm the samples to room temperature (15–30°C).

    b) To a 25 mL filtrate sample (from 1(b) above), add 2.5 mL of mixed reagent and mix at once.

    c) After 5 min, and preferably within the first 2–3 h, measure the extinction in a 10-cm cell against distilled water at 885 nm.

    d) Use distilled water in place of a sample and carry out steps (a) to (c) above to obtain the extinction of the reagent blank. (Reagent blanks should be less than 0.02 on a 10-cm cell).

    e) Determine the factor (it is used to convert extinction or absorbance into μM concentration) for a 3 μM PO$_4^{-3}$ standard as follows: Dilute 10 mL of the standard to 1 L with distilled water. Pipette 1.25 mL

of dilute standard into each of three Erlenmeyer flasks and make up
to 25 mL with distilled water from a graduated cylinder. Carry out
the procedure in steps (a) to (c). Calculate the factor, F, as:

$$F = \frac{3.00}{E_S - E_B}$$

where E_S is the average extinction of triplicate standards and E_B is
the average extinction of the reagent blank. The value of F should
be about 5.

f) Correct the extinction of the sample with the reagent blank and calculate the phosphate concentration as:

$$\mu M = \text{corrected extinction} \times F$$

where F is the factor as described in (e).

Calculations

Calculate the uptake rate (V) for each culture as follows:

$$V = \frac{C_0 - C_1}{N \cdot t}$$

C_0 & C_1 = concentration of PO_4 at beginning
and end of time period (μM)

N = cell density (cells L^{-1})

t = time period of uptake (h)

V = uptake rate (μmol 10^6 cells^{-1} h^{-1})

Questions

1. a) Compare the maximal uptake rates of phosphate for the P-saturated
   and P-starved cultures. Why are they different?
   b) Compare your results to other work in the literature on the effects
   of nutritional history on nutrient uptake rates (see references).
2. What is the ecological importance of the rapid uptake potential in
   nutrient limited cells?
3. Could this method be used to detect phosphate limitation in phytoplankton? Explain.

REFERENCES

Aitchinson, P. A. & V. S. Butt (1983). The relationship between the synthesis of
inorganic polyphosphate and phosphate uptake by *Chlorella vulgaris*. *J. Exp.
Bot.* 24, 497–510.

Cembella, A. D., N. J. Antia & P. J. Harrison (1984a). The utilization of inorganic and organic phosphorus compounds as nutrients by eukaryotic microalgae: a multidisciplinary perspective: Part I. *Crit. Rev. Microbiol.* 10, 317–391.

Cembella, A. D., N. J. Antia & P. J. Harrison (1984b). The utilization of inorganic and organic phosphorus compounds as nutrients by eukaryotic microalgae: a multidisciplinary perspective: Part II. *Crit. Rev. Microbiol.* 11, 13–81.

Collos, Y. (1983). Transient situations in nitrate assimilation by marine diatoms, 4. Non-linear phenomena and the estimation of the maximum uptake rate. *J. Plankton Res.* 5, 667–691.

Conway, H. L., P. J. Harrison & C. O. Davis (1976). Marine diatoms grown in chemostats under silicate or ammonium limitation. II. Transient response of *Skeletonema costatum* to a single addition of the limiting nutrient. *Mar. Biol.* 35, 187–199.

Dowd, J. E. & D. S. Riggs (1965). A comparison of estimates of Michaelis-Menten kinetic constants from various linear transformations. *J. Biol. Chem.* 240, 863–869.

Droop, M. R. (1983). Twenty-five years of algal growth kinetics. *Bot. Mar.* 26, 99–112.

Harrison, P. J. & L. D. Druehl (1982). Nutrient uptake and growth in the Laminariales and other macrophytes: a consideration of methods. In, L. M. Srivastava (ed.) *Synthetic and Degradative Processes in Marine Macrophytes.* Walter de Gruyter, Berlin, pp. 99–120.

Harrison, P. J., D. H. Turpin, J. S. Parslow & H. L. Conway. In Press. Determination of nutrient uptake kinetic parameters: A comparison of methods. *J. Exp. Mar. Biol. Ecol.*

Healey, F. P. (1980). Slope of the Monod equation as an indicator of advantage in nutrient competition. *Microbiol. Ecol.* 5, 281–286.

Healey, F.P. (1978). Phosphate uptake. In, J. A. Hellebust & J. S. Craigie (eds.) *Handbook of Phycological Methods: Physiological and Biochemical Methods.* Cambridge University Press, New York, pp. 412–417.

Lean, D. S. R. & E. White (1983). Chemical and radiotracer measurements of phosphorus uptake by lake plankton. *Can. J. Fish. Aquat. Sci.* 40, 147–155.

Nalewajko, C. & D. R. S. Lean (1980). Phosphorus. In, I. Morris (ed.) *The Physiological Ecology of Phytoplankton.* University of California Press, Berkeley, pp. 235–258.

McCarthy, J. J. (1981). The kinetics of nutrient utilization, In, T. Platt (ed.) *Physiological Bases of Phytoplankton Ecology, Can. Bull. Fish. Aquat. Sci.* pp. 210, 211–233.

Parslow, J. S., P. J. Harrison & P. A. Thompson (1984). Saturated uptake kinetics: transient response of the marine diatom *Thalassiosira pseudonana* to ammonium, nitrate, silicate or phosphate starvation. *Mar. Biol.* 83, 51–59.

Parsons, T. R., Y. Maita & C. M. Lalli (1984). *A Manual of Chemical and Biological Methods for Seawater Analysis.* Pergamon Press, New York, 123 pp.

Perry, M. J. (1976). Phosphate utilization by an oceanic diatom in phosphorus-limited chemostat culture and in the oligotrophic waters of the central North Pacific. *Limnol. Oceanogr.* 21, 88–107.

Strickland, J. D. H. & T. R. Parsons (1972). *A Practical Handbook of Seawater Analysis. Fish. Res. Bd. Canada Bull.*, 167, 310 pp.

NOTES FOR INSTRUCTORS
Cultures

Several fast growing diatoms can be used (e.g., *Thalassiosira pseudonana, Phaeodactylum tricornutum, Skeletonema costatum*). Some flagellates such as *Dunaliella tertiolecta* do not exhibit enhanced uptake rates and large cells respond slowly, so the results will not be as evident. Culture density should be $>10^8$ cells L^{-1}, otherwise the amount of nutrient taken up in one hour will be too small to measure. Culture densities near 10^9 cells mL^{-1} may be required for the P-saturated culture, because the uptake rate will be much slower than the P-deficient culture.

Medium and culturing. Use autoclaved, filtered natural seawater and enrich with N, P and Si so that the final concentrations are 100 μM NO_3^-, 1 μM PO_4^{-3} and 100 μM SiO_4^{-4}. Add trace metals and vitamins to about f/2 (Guillard and Ryther 1962), ES/2 (Harrison et al. 1980), or PES (Exp. 28). Monitor culture growth using *in vivo* fluorescence (see Exp. 4B). On the day of the experiment, the fluorescence of the P-starved culture should have *begun* to plateau (indicating P-starvation; 24 to 48 h starvation is optimal), while the P-saturated culture should still be in logarithmic phase growth. If cultures are ready too early, simply dilute the cultures (the amount will have to be calculated and based on their previous growth rate and the amount of time remaining until the start of the experiment). Artificial seawater may also be used (Harrison et al. 1980; Morel et al. 1979).

Lights and temperature. Use 100 μE m^{-2} s^{-1} and fluorescent lights (cool white or daylight). Incubate at 18°–20°C in a constant temperature water bath or temperature-controlled growth chamber.

Materials

1. Phytoplankton cultures (P-starved and P-saturated) with cell density $>$ 10^8 cells L^{-1} (100 mL) (see earlier discussion)
2. 0–1 and 0–5 mL adjustable pipettes or volumetric pipettes
3. membrane filters (Millipore; 0.45 to 1 μm pore size) or glass fiber filters (Gelman or Whatman)
4. filtration apparatus
5. Palmer-Maloney chamber or Neubauer haemocytometer for cell counts
6. Erlenmeyer flasks (50 mL) or large test tubes (50 mL)

Instruments

1. microscope for counting phytoplankton
2. spectrophotometer
3. vacuum pump

Chemicals and reagents

1. PO_4^{-3} stock solution (1 μmol mL^{-1}) (100 mL)
2. Phosphate Primary Standard:
   Dissolve 0.816 g of anhydrous potassium dihydrogen phosphate, KH_2PO_4, in 1 L of distilled water. Store in a dark bottle with 1 mL of chloroform; the solution is stable for many months.
3. Ammonium molybdate solution:
   Dissolve 15 g of analytical reagent grade ammonium paramolybdate $(NH_4)_6Mo_7O_{24} \cdot 4H_2O$ in 500 mL of distilled water. Store in plastic bottle away from direct sunlight. The solution is stable, but if high blank values are obtained, this solution should be made up again and tested.
4. Sulfuric acid solution:
   Slowly add 140 mL of concentrated (sp. gr. 1.82) analytical reagent quality sulfuric acid to 900 mL of distilled water. Allow the solution to cool and store it in a glass bottle.
5. Ascorbic acid solution:
   Dissolve 27 g of ascorbic acid in 500 mL of distilled water. Store the solution in a plastic bottle frozen solid in the freezer. The solution is stable for many months but should not be kept at room temperature for more than a week.
6. Potassium antimonyl-tartrate solution:
   Dissolve 0.34 g of potassium antimonyl-tartrate (tartar emetic) in 250 mL of water, warming if necessary. Store in a glass or plastic bottle. The solution is stable for many months.
7. Mixed reagent:
   Mix together 20 mL ammonium molybdate, 50 mL sulfuric acid, 20 mL ascorbic acid, and 10 mL of potassium antimonyl-tartrate solutions. Prepare this reagent when needed (store no more than 6 h) and discard any excess. Do not store; the quantity is suitable for about 40 samples.

Calculation of kinetic parameters

The hyperbolic equation $V = V_{max} S/(K_s + S)$ can be rearranged or transformed three ways, producing three different linear plots. They are:

1. $1/S$ vs. $1/V$ (Lineweaver–Burk plot), slope $= K_s/V_{max}$ and y-intercept $= V_{max}$.
2. S/V vs. S (Woolf plot), slope $= 1/V_{max}$ and y-intercept $= K_s/V_{max}$.
3. V/S vs. V (Eddie-Hofstee plot), slope $= -K_s$ and y-intercept $= V_{max}$.

The most commonly used plot for algal uptake kinetics is the Woolf plot. The hyperbolic equation is linearized as follows:

$$V = V_{max} \cdot S/(K_s + S)$$
$$V_{max} \cdot S = V(K_s + S)$$
$$S/V = 1/V_{max}(K_s + S)$$
$$S/V = K_s/V_{max} + S/V_{max}$$

Compare to the general equation for a straight line $y = a + bx$ where $a =$ slope and $b =$ y-intercept.

Problems associated with phosphate analysis

If no spectrophotometer capable of reading 885 nm is available, readings can be made at a somewhat lower wavelength (e.g. 850 nm) with some loss of sensitivity (Strickland & Parsons 1972).

The method described in this experiment measures "soluble reactive phosphorus," or "SRP". Soluble refers to phosphorus that passes through a 0.45 μm filter, and reactive refers to phosphorus compounds or ions that react with the reagents used in the phosphate method. However, the acidic nature of the reagents cause hydrolysis of organic phosphorus esters. These phosphate ions that are liberated cause the estimation of inorganic phosphorus concentrations to be overestimated, especially if color development time is long (>0.5 h). For further discussion of the problems of phosphate analysis, see Cembella et al. 1984a (p. 319–24). One of the best ways to measure phosphate is with radiochemical analysis using ^{32}P (Levine & Schindler, 1980). This technique is tedious and requires considerable expertise.

Additional References

Guillard, R. R. L. & J. H. Ryther (1962). Studies of marine planktonic diatoms. I. *Cyclotella nana* Hustedt and *Detonula confervacea* (Cleve) Gran. *Can. J. Microbiol.* 8, 229–239.

Harrison, P. J., R. E. Waters & F. J. R. Taylor (1980). A broad spectrum artificial seawater medium for coastal and open ocean phytoplankton. *J. Phycol.* 16, 28–35.

Morel, F. M. M., J. G. Reuter, D. M. Anderson & R. R. L. Guillard (1979). Aquil: A chemically defined phytoplankton culture medium for trace metal studies. *J. Phycol.* 15, 135–141.

Levine, S. N. & D. W. Schindler (1980). Radiochemical analysis of orthophosphate concentrations and seasonal changes in the flux of orthophosphate to seston in two Canadian Shield Lakes. *Can. J. Fish. Aquat. Sci.* 37, 472–478.

22

Nitrogen metabolism and measurement of nitrate reductase activity

David J. Chapman

Dept. of Biology, University of California, Los Angeles, CA
90024 USA

Paul J. Harrison

Dept. of Oceanography, University of British Columbia,
Vancouver, BC, Canada, V6T 2B1

FOR THE STUDENT

Introduction

Nitrogen-containing compounds of organisms (e.g., proteins, amino acids, purine and pyrimidine bases) have nitrogen in the reduced (-3 valency) state. In most unpolluted natural waters (e.g., rivers, lakes, open oceans) available inorganic nitrogen is usually present in the oxidized ($+5$ valency) state, as nitrate.

Organisms that assimilate nitrate (frequently the sole source of nitrogen) must therefore reduce the nitrate to "ammonia" or amino form for incorporation into organic compounds. This assimilatory nitrate reduction proceeds via a number of individual reactions. One of the most important of these is the first step in the sequence: the reduction of nitrate to nitrite by the enzyme nitrate reductase

$$NO_3^- + 2H^+ + 2e^- \rightarrow NO_2^- + H_2O$$

The source of the reductant (electron donor) is actually NADH; in some cases NADPH (e.g., some higher plants). The enzyme activity in cell extracts can be readily monitored by measuring the appearance of the product nitrite NO_2^-.

This enzyme has been assayed in a number of micro-algae (Morris & Syrett 1963; Rigano & Violante 1973; Hersey & Swift 1976) and macroscopic algae (Davison & Stewart 1984). The assay procedure is based upon determining the rate of reduction of nitrate to nitrite (Eppley et al. 1969). The underlying analytical chemistry of the assay is also the same as that used for determining nitrate in natural waters (Grasshoff et al. 1983; Strickland & Parsons 1972), except in that case the nitrate is reduced to nitrite by passage through a Cd-Cu column.

The purpose of this exercise is to assay for nitrate reductase in a marine macrophytic alga or a single-celled green alga. An additional experiment involves the determination of the effect, if any, of the ammonium ion on the *in vivo* presence of the enzyme.

Procedure: Nitrate Reductase Activity

You will be provided with fresh material of an appropriate marine macrophyte or cells of *Chlorella pyrenoidosa* grown on a synthetic nitrate-containing medium, and harvested in late log phase of growth (14 days).

1. If you use a macroalga, quickly rinse with filtered seawater. If you use cultured *Chlorella*, wash the cells three times by centrifugation with distilled water.

Extraction procedure if you assay the Chlorella

2. Weigh a weighing dish, bottle or centrifuge tube with aluminum foil cover. Add and weigh an aliquot of your *Chlorella* cells for dry weight determination. Determine the dry weight of the aliquot by drying the alga for 2 h in an oven at 110°C.
3. Weigh the remaining cells and place in a chilled mortar. There should be about 2 g of cells. Add ca. 1.5 g grinding sand or alumina and grind vigorously with pestle while adding in stages 25 mL of ice cold buffer (0.2 M potassium phosphate buffer, pH 8.5 with 1 mM dithiothreitol).

Extraction procedure if you assay the macrophytic alga

2. Weigh a weighing dish, bottle or centrifuge tube with aluminum foil cover. Add and weigh an aliquot of your alga for dry weight determination. Determine the dry weight of the sample by drying the alga for 2 h in an oven at 110°C.
3. Weigh the remaining tissue and place in a chilled mortar. There should

be about 2 g of tissue. Add ca. 1.5 g grinding sand or alumina and
0.1 g polyvinyl polypyrrolidone (PVPP) (to bind with phenolic compounds that are contained in some macrophytes and inhibit enzyme
activity). Grind vigorously with pestle while adding in stages 25 mL of
ice cold buffer (0.2 M potassium phosphate buffer, pH 8.5 with 1
mM dithiothreitol).

Enzyme assay for all extracts

4. Centrifuge the extract in a refrigerated centrifuge for 15 min at 15,000
   $\times$ g. Recover the supernatant, which should be clear.
5. To 4 mL of the supernatant in a centrifuge tube, add 0.5 mL of 0.33
   M potassium phosphate buffer, pH 8.5; 0.2 mL of 0.2 M KNO$_3$; 0.1
   mL of 2 mM NADH; 0.1 mL of 0.05 M MgSO$_4$·7H$_2$O.
6. Prepare a control mixture in a separate test tube with the 0.1 mL of
   NADH replaced by 0.1 mL of potassium phosphate buffer.
7. Mix the reactants well. Cover both control and sample tubes with
   aluminum foil or a small beaker and place both in a water bath at
   25°C. Incubate for 30 min.
8. Stop the reaction by adding 0.2 mL of 0.1 M zinc acetate and 3.4 mL
   of ethanol. Shake thoroughly, centrifuge (refrigerated centrifuge,
   10,000 $\times$ g, 10 min) and decant the supernatants into fresh tubes.
9. To each of the two supernatants (control and sample) add 0.5 mL of
   4% sulfanilimide in 3N HCl and 1.0 mL of 0.1% N(1-naphthyl) eth-
   ylene-diamine HCl in H$_2$O. Cover and shake well. A red violet color
   will appear. Allow 15 min for the reaction to complete.
10. Record the absorbance at 543 nm and subtract the absorbance of the
    control. The color formed by the nitrite complex is stable for several
    minutes.
11. From the standard graph provided, calculate the concentration of
    nitrite in your sample.
12. Express and tabulate your results as μmol nitrite formed per gram dry
    weight per unit time.

Procedure: effect of nitrogen source on nitrate reductase activity

You will be provided with cells of *Chlorella pyrenoidosa* grown under
each of the following three conditions.

1. Grown in medium with KNO$_3$ as nitrogen source for 2 weeks.
2. Grown in medium with KNO$_3$ as nitrogren source for 12 days and then
   maintained in nitrogen-free medium for 2 days.
3. Grown in medium with KNO$_3$ as nitrogen source for 12 days and then
   transferred to medium with NH$_4$Cl as nitrogen source for 2 days.

Measure nitrate reductase activity in each of the three samples as outlined in the preceding Steps 1–11. Tabulate your results as relative enzyme activity, with KNO_3 culture expressed as 100.

Questions

1. Why is the nitrate assay based upon a nitrite reaction and not a nitrate reaction?
2. What effect does a two-day exposure to an NH_4^+ sole source nitrogen have on enzyme activity?
3. What effect does a nitrogen-free medium have on enzyme activity?
4. What are possible sources of error?
5. What is the role of NADH in the reaction mixture?
6. How does zinc acetate-ethanol stop the reaction?
7. What is the purpose of the dithiothreitol?
8. What assumptions have been made in the design and execution of this experiment?

REFERENCES

Davison, I. R. & W. D. P. Stewart (1984). Studies on nitrate reductase activity in *Laminaria digitata* (Huds.) Lamour. I. Longitudinal and transverse profiles of nitrate reductase activity within the thallus. *J. Exptl. Mar. Biol. Ecol.* 74, 201–210.

Eppley, R. W. (1978). Nitrate reductase in marine phytoplankton. In, J. A. Hellebust, and J. S. Craigie (eds) *Handbook of Phycological Methods. Physiological and Biochemical Methods.* Cambridge University Press, Cambridge, pp. 217–223.

Eppley, R. W., J. L. Coatsworth & L. Solórzano (1969). Studies of nitrate reductase in marine phytoplankton. *Limnol. Oceanogr.* 14, 194–205.

Grasshoff, K., M. Ehrhardt & K. Kremler (1983). *Methods of Seawater Analysis.* and ed., Verlag Chemie. Berlin, 419 pp.

Hersey, R. L. & E. Swift (1976). Nitrate reductase activity of *Amphidinium carteri* and *Cachonina niei* (Dinophyceae) in batch culture: Diel periodicity and effects of light intensity and ammonia. *J. Phycol.* 12, 36–44.

Morris, I. & P. J. Syrett (1963). Development of nitrate reductase in *Chlorella* and its repression by ammonia. *Arch. Mikrobiol.* 47, 32–44.

Rigano, C. & U. Violante (1973). Effect of nitrate, ammonia and nitrogen starvation on the regulation of nitrate reductase in *Cyanidium caldarium. Arch. Mikrobiol.* 90, 27–33.

Strickland, J. D. H. & T. R. Parsons (1972). A practical handbook of seawater analysis. 2nd Ed. *Bull. Fish. Res. Bd. Canada.* Number 167. 311 pp.

NOTES FOR INSTRUCTORS

It is suggested that each pair of students conduct one assay.

Supplies for each pair of students

1. clinical centrifuge tubes (12 mL) (12)
2. test tube rack with 12 test tubes
3. serological pipettes (1 mL) (12)
4. serological pipettes (5 mL) (6)
5. serological pipettes (10 mL) (6)
6. pipette bulbs (6)
7. small pestle and mortar
8. aluminum foil or small beakers (to cover test tubes)
9. grinding sand or alumina—ca. 20 g.
10. weighing bottles
11. polyvinyl polypyrrolidone, ca. 2 g (for macroalgae only)

Reagent solutions
(for entire class of 24)

All reagent solutions should be prepared fresh the day before and stored in the refrigerator.

1. 1000 mL 0.2 M potassium phosphate buffer pH 8.5 containing 1 mM dithiothreitol
2. 100 mL 0.33 M potassium phosphate buffer pH 8.5
3. 100 mL 0.2 M KNO_3 solution
4. 50 mL 2 mM NADH
5. 100 mL 0.05 M $MgSO_4 \cdot 7H_2O$ solution
6. 100 mL 1 M zinc acetate solution
7. 500 mL 100% ethanol
8. 100 mL 4% sulfanilimide in 3N HCl
9. 100 mL 0.1% N(1-naphthyl) ethylene diamine HCl
10. 1000 mL 5 mM $NaNO_2$ *(nitrite)* (for preparing standard curve)
    NADH, Sulfanilimide, N(1-naphthyl) ethylene diamine available from Sigma Chemical Co. or other organic chemical supply companies.

Instrumentation

1. analytical balance
2. clinical centrifuges
3. water bath
4. spectrophotometer with 1 mL cuvettes
5. refrigerated centrifuge with 250 mL bottles and 40 mL and 15 mL tubes
6. drying oven

Cultures

Culture of *Chlorella pyrenoidosa* UTEX 251 available from University of Texas Culture Collection of Algae.

Base medium—nitrogen free.

NaCl	450 mg
K_2HPO_4	175 mg
KH_2PO_4	400 mg
$CaCl_2 \cdot 2H_2O$	150 mg
$MgSO_4 \cdot 7H_2O$	250 mg
Fe-EDTA solution	1 mL*
Trace element solution	1 mL**
H_2O to 1000 mL	

Fe-EDTA solution

$$4 \text{ g FeSO}_4 \cdot 7H_2O + 5 \text{ g Na}_2EDTA \text{ in } 500 \text{ mL } H_2O$$

**Trace Element Solution (Arnon's)*

H_3BO_3	2.99 g
$MnCl_2 \cdot 4H_2O$	1.8 g
$ZnSO_4 \cdot 7H_2O$	2.0 g
$CuSO_4 \cdot 5H_2O$	80 mg
Ammonium paramolybdate	180 mg
$CoCl_2 \cdot 6H_2O$	40mg
H_2O to 1000 mL	

For the nitrate medium add 50 mg L^{-1} KNO_3.

For the ammonium medium add 26 mg L^{-1} NH_4Cl. Adjust to pH 7.8.

For each three student pairs you will need about 3000 mL of culture. This is obtained by inoculating into 3000 mL of nitrate medium ca. 300 mL of nitrate grown cells 14 days prior to the class. Cells are grown at 18°–20°C at ca. 150 μE m^{-2} s^{-1} irradiance. Occasionally shake cultures. Aeration is helpful but not essential.

To set up the nitrogen-free and ammonium cultures, take 2000 mL of culture at 12 days. Gently centrifuge the cells in a refrigerated centrifuge (ca 15°–20°C, 10 min at 3000 × g). Thoroughly resuspend the cells in nitrate-free medium and recentrifuge. Repeat three times to ensure the removal of nitrate. Resuspend half the cells in 1000 mL ammonium medium and half in 1000 mL nitrogen-free medium. Maintain in culture for two more days as described earlier. Since growth is light and temperature dependent, growth should be monitored by *in vivo* chlorophyll fluorescence. This procedure will give the following:

1000 mL nitrate grown culture 14 days
1000 mL nitrate grown culture 12 days, followed by two days NH_4^+
1000 mL nitrate grown culture 12 days, followed by two days nitrogen free.

This will provide for three pairs, each pair conducting one assay. Cells are harvested just prior to experiment by centrifugation in refrigerated centrifuge; (10 min at 10,000 $\times$ g), and the pellet stored in the tube in ice until the start of the class.

The *Porphyra* or other macrophytic marine alga should ideally be collected the day of the experiment and maintained in cooled seawater, or collected and maintained two to three days in running seawater in room light.

Do not use macroalgae such as *Fucus, Pelvetiopsis,* and *Egregia* that contain large amounts of phenolics and/or polysaccharides. Select young active tissue (new spring growth) if possible.

It is advisable to do a test assay prior to the laboratory. Amounts and activity may vary, and the class may need to dilute the original cell extract, or use smaller aliquots of material.

Preparation of the standard nitrite curve

Follow Steps 4, 5, 7, and 8 of the procedures for students, but replace the 4 mL of supernatant with 4 mL of $NaNO_2$ solution, in each of seven samples:

4 mL of 1 mM $NaNO_2$ · Final NO_2^- is 0.4 mM
4 mL of 0.5 mM $NaNO_2$ · Final NO_2^- is 0.2 mM
4 mL of 0.25 mM $NaNO_2$ · Final NO_2^- is 0.1 mM
4 mL of 0.1 mM $NaNO_2$ · Final NO_2^- is 0.04 mM
4 mL of 0.05 mM $NaNO_2$ · Final NO_2^- is 0.02 mM
4 mL of 0.025 mM $NaNO_2$ · Final NO_2^- is 0.01 mM
4 mL of 0.01 mM $NaNO_2$ · Final NO_2^- is 0.004 mM

4 mL of H_2O. Final NO_2^- is zero. This is the control blank.

Follow Step 9 to record the absorbance. Graphically plot NO_2^- concentration against Absorbance.

23

Turgor pressure regulation in marine macroalgae

G. O. *Kirst*

Fachbereich 2 (Biologie-Chemie) - Botanik, University of
Bremen, D-2800 Bremen, FRG

FOR THE STUDENT

Introduction

Growth and distribution of algae are primarily controlled by light, temperature, nutrients and salinity. Salinity is highly variable in coastal regions, especially in the intertidal zones, estuaries, and rock pools. Freshwater during rain periods can decrease salinity markedly or the salt concentrations may more than double by evaporation of water in shallow rock pools or littoral habitats during low tide. Many algae growing in these habitats tolerate changes in salinity by maintaining a constant turgor pressure. However, not only these species, but also those abundant in the more constant habitats of deep waters maintain constant turgor pressure when subjected to changes in the external salinity.

Turgor pressure (P) arises from the osmotic gradient between the inside and the outside of a cell. Thus, changing external osmotic pressure (π_o, proportional to salinity) will result in changes of turgor pressure and volume of the algal cell. Because turgor pressure is the driving force behind growth and imparts stability and shape to the whole plant, changes in turgor pressure are tolerated only transiently. Disturbance of turgor pressure will affect metabolism via changes in cell water potential and spatial orientation of membrane systems, e.g., thylakoids of chloroplasts. Restoration of the original status is essential. The adaptation or "osmotic adjustment" is complex and involves (1) sensing the external change of salinity,

203

(2) changes in metabolism and/or ion transport, and (3) regulation of the reactions mentioned under (2), depending on the degree of adjustment achieved.

Two phases of response can be observed in algal cells subjected to an osmotic stress (hyperosmotic: increasing salinity; hypo-osmotic: decreasing salinity). The first phase is characterized by changes in turgor pressure or (in wall-less cells) cell volume, caused by rapid water fluxes following the osmotic gradient. This phase is very fast (5–20 sec in microalgae, min in macroalgae). The second phase results in restoration of turgor pressure or cell volume due to metabolism (synthesis or degradation of low molecular weight organic compounds) and ion transport (stimulation or inhibition of selective ion uptake, e.g., K^+, Cl^-). This phase is comparatively slow and takes place over 1–2 h for microalgae and 24–72 h for macroalgae.

Under steady state conditions, turgor pressure (P) equals the difference ($\Delta\pi$) of internal (π_i) and external (π_o) osmotic potential. Hence, if we estimate π_o and π_i we can calculate turgor pressure simply by

$$\text{Equation 1: } P \equiv \Delta\pi = \pi_i - \pi_o$$

(see Gutknecht et al. 1978; Nobel 1982; Lobban et al. 1985: pp. 50–53). In dilute solutions, osmotic potential is one of the four colligative properties, which depend on the concentration of solutes regardless of their nature. Measuring another of the colligative properties—freezing point depression, boiling point elevation or vapor pressure reduction—is a direct and convenient way of estimating osmotic potential. Most modern, half-automated laboratory osmometers use either freezing point depression (minimum amount of sample ca. 30–200 μL) or vapor pressure reduction (amount of sample 5 μL).

A manual cryoscopic method (measuring freezing point depression) is used here to estimate osmotic potential of saps expressed from algal cells or tissues. The sap samples are obtained from portions of algal thalli or single giant cells, which have been adapted for four to six days to media of various salinities. Different species will exhibit different turgor pressures (Table 23.1). Over a range of salinities the turgor pressure in each species should not differ, indicating that the alga can regulate its osmotic potential to maintain a steady state.

Procedure

Harvesting. Draw samples from the experimental media to estimate π_o. Blot the algae on a paper tissue to remove adhering medium.

Table 23.1. *Turgor pressure of some marine (m)*
and freshwater (f) algae (1 MPa = 10 bar).

Species	Turgor pressure MPa
Chlorophyta	
Chaetomorpha linum (m)	1.0–1.7
Valonia utricularis (m)	0.15–0.2
Caulerpa racemosa (m)	0.08–0.1
Halicystis parvula (m)	0.05
Chara fragilis (f)	0.6–0.8
Nitella flexilis (f)	0.5–0.8
Phaeophyta	
Scytosiphon lomentaria (m)	0.5–0.62
Ecklonia radiata (m)	0.23–0.28
Rhodophyta	
Griffithsia monilis (m)	0.25–0.34
Lomentaria umbellata (m)	0.13–0.2

Giant cells (*Valonia* sp.; cells of brackish water Characeae etc.): Place cells on a wax plate and cut open with a razor-blade. Collect the vacuolar sap with a syringe or pipette and transfer to a stoppered and labelled vial.

Other algae: Mince tissue with scissors and press through a garlic press. Alternatively: Cut tissue in small pieces into a centrifuge tube, mash with a glass rod and centrifuge at maximum speed. Expressed saps can be stored for several weeks at −20°C, tightly sealed.

Measurement of osmotic potentials

1. In two 400 mL beakers a cooling mixture is prepared. This consists of 35 g NaCl plus 200 mL tap water, to which ice is added up to the 400 mL mark. The mixture is stirred and the temperature should be checked to see that it is approximately −6°C. Place the cooling jacket into one of the beakers (see Fig. 23.1).
2. Put sufficient of the sap sample (0.2–0.5 mL) into the sample container to cover the mercury bulb of the thermometer and put in the stopper.
3. For precooling, place the sample container prepared according to Step 2, into the ice-bath of the second beaker. The sample has to be cooled below freezing (−2 to −3°C). The sample must not be stirred or shaken or premature ice formation will take place. Now stir gently to initiate ice formation by scratching the glass container with the sample mixer. The sample will freeze and the temperature rises.

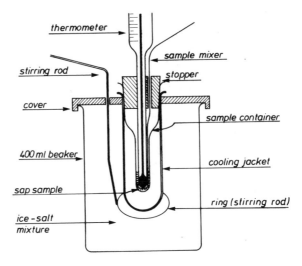

Fig. 23.1. Cryoscope for estimating freezing point depression to calculate the osmotic potential of solutions. See materials for list of components.

4. Insert sample container into the cooling jacket and stir carefully until the temperature remains constant. Read temperature as "uncorrected freezing point" (FD1).
5. Thaw the sample by holding the sample container tightly in the hand and repeat measurements according to Steps 3 and 4. Calculate the average of three measurements.
6. Repeat procedure 2–5, with the media as samples to estimate π_o.

Calculations

A concentration of 1 mol of an undissociated solute in 1000 g water (1 osmol kg^{-1}) results in a freezing point depression of 1.86°C. At 0°C the osmotic potential, π, of a one molal solution equals 2.27 MPa (mega Pascal: 10^6 Pa = 10 bar = 9.87 atm).

Hence, the equations are:

$$\text{Equation 2: } \pi_{0°C} = \frac{2.27\ FD}{1.86} = 1.22FD\ \text{MPa}$$

or at 20°C:

$$\text{Equation 3: } \pi_{20°C} = 1.39FD\ \text{MPa}$$

For comparison with the data from literature, the following relationships may be useful (20°C):

$$1000 \text{ m osmol kg}^{-1} = 2.58 \text{ MPa} = 25.85 \text{ bar} = 22.5 \text{ atm.}$$

Data should be presented in a table including uncorrected freezing point depression (FD1), true freezing point depression (FD), π_i of cell sap (20°C) and π_o (20°C) of the media, $\Delta\pi$ ($\equiv$ turgor pressure; equation 1). Draw a graph of $\Delta\pi$ versus π_o.

Questions

1. Consider a wall-less cell (e.g., a unicellular flagellate such as *Dunaliella* sp): What could you observe with respect to the cell volume after hyperosmotic or hypo–osmotic shocks.
2. Explain what happens physiologically during plasmolysis.
3. The marine giant-celled algae *Valonia* regulates turgor pressure by changing its internal K^+ concentration. What would you expect with hypo-osmotic conditions considering an active K^+ uptake mechanism?
4. *Valonia* cells are big enough to be impaled with probes of glass capillaries. If a small droplet of incompressible oil is pressed into the vacuole while external salinity is held constant, what will happen immediately to turgor pressure and to K^+ uptake?
5. A 0.01 M solution of sucrose has, to a good approximation (dilute solution!), a π of 10 mosmol kg^{-1}. What would be approximately the π of a 0.01 M NaCl solution? a 0.01 M Na$_2$SO$_4$ solution?

REFERENCES

Gutknecht, J., D. F. Hastings & M. A. Bisson (1978). Ion transport and turgor pressure regulation in giant algal cells, In, G. Giebisch, D. C. Tosteson & H. H. Ussing (eds) *Membrane Transport in Biology: III Transport Across Multimembrane Systems*. Springer Verlag, Berlin, pp. 125–174.

Hellebust, J. A. (1976). Osmoregulation. *Ann. Rev. Plant Physiol.* 27, 485–505.

Kirst, G. O. & M. A. Bisson (1979). Regulation of turgor pressure in marine algae: Ions and low-molecular weight organic compounds. *Aust. J. Plant Physiol.* 6, 539–556.

Lobban, C.S., P. J. Harrison & M. J. Duncan (1985). *The Physiological Ecology of Seaweeds*. Cambridge University Press, New York.

Nobel, P. S. (1982). *Biophysical Plant Physiology*. W. H. Freeman and Company, San Francisco.

NOTES FOR INSTRUCTORS
Materials

Cryoscopes. Two cryoscopes as shown in Fig. 23.1 (or a lab osmometer, e.g., Wescor 5100B vapor pressure osmometer). Each cryoscope consists of:

1. 400 mL beakers (2) (one for precooling)
2. stirring rod: heavy wire to mix the ice bath (1)
3. cover with holes for the stirring rod and to hold the cooling jacket; made out of aluminum sheet, plastic or wood (1)
4. cooling jacket: appropriate test tube (ca. 3 cm diameter) (1)
5. sample mixer: thin hooked wire, stainless steel (1)
6. thermometer with long bulb (1)
7. sample container: Glass tube with a wide top, fitting into the cooling jacket. The lower part should be narrowed to allow space only for the thermometerbulb, the sample mixer, and as little sample volume as possible, just enough to cover the mercury bulb of the thermometer (0.2–0.5 mL).

Other equipment

1. Erlenmeyer flasks 500 mL (12) (3 for each salinity)
2. volumetric flasks or measuring cylinders 1L (2)
3. pipettes 2 mL (4)
4. syringes 1 mL (4)
5. scissors (2)
6. razor-blades (2)
7. wax plate (1)
8. centrifuge tubes 10 mL (8)
9. lab centrifuge (1)
10. garlic press or mortar with pestle (1)
11. shaker for the Erlenmeyer flasks or air pump for aeration (1)
12. light source (1)

Chemicals. NaCl; ice; seawater (filtered) or artificial seawater.

Algae. All giant-celled or coenocytic algae that contain little or (better) no mucilage are suitable: e.g., *Valonia* spp.; *Dictyosphaeria* spp.; *Chaetomorpha aerea*, (*C. melagonium*, *C. darwinii*); *Bryopsis* spp.; *Halicystis* spp.; *Griffithsia monilis*; and brackish water Charophytes. *Caulerpa* spp., *Codium* spp., kelps, and most other brown and red algae are not recommended unless a vapor pressure osmometer (e.g., Wescor) is available. This is because of the high viscosity of the expressed saps. Their advantage, however, is that the thallus can be cut in pieces. After being given a day for healing, they yield enough homogenous plant material for the experiment. Algae should be cleaned carefully after collection to remove epiphytes.

Osmotic adaptation

100 mL medium per 2–5 g fresh weight of tissue is used. Different salinities are obtained by dilution of filtered seawater (or artificial seawater medium) with solutions of 2 mM NaHCO$_3$ to yield a 0.5 $\times$ seawater (hypo-osmotic conditions) or by addition of NaCl to yield a 1.5 $\times$ and a 2 $\times$ media (hypersmotic conditions). Normal seawater strength (35‰) is used as control. The media together with the algal samples are put in labelled Erlenmeyer flasks, plugged with cotton and left under continuous low illimination (50–100 μE m^{-2} s^{-1}) for four to six days. Irradiation with the usual artificial light is too bright; two to three tubes (20 Watt) of cool white light in a distance of 0.5–0.75 m is sufficient. The flasks can be shaken gently on a shaker or, alternatively, mixing can be achieved by bubbling with air. It is important to pass the air through a water-filled washing bottle to moisten it.

Calibration

The thermometer must be calibrated for this method because the volume and heat-storing capacity of the mercury bulb are large compared to the sample and affect the readings. The correction factor (FD-H$_2$O) is estimated using distilled water and proceeding as in Steps 2–5 above. The true freezing point depression (FD) of the sap sample is calculated as the difference between the uncorrected freezing point (FD1) and FD-H$_2$O.

$$FD = FD1 - FD\text{-}H_2O.$$

To save time, the correction factors (FD-H$_2$O) for each of the thermometers used should be estimated separately and provided by the instructor.

Scheduling

Four to six days osmotic adaptation. Harvesting, preparation of expressed saps and cryoscopic estimation of osmotic potential can be done within a 2–3 h laboratory period using two cryoscopes: four measurements each, a control (seawater) plus one hypo–osmotic and two hyperosmotic conditions.

24

Organic osmotica: their role as "compatible solutes" in response to salinity

G. O. Kirst

Fachbereich 2 (Biologie-Chemie) - Botanik, University of
Bremen, D-2800 (FRG)

FOR THE STUDENT

Introduction

Enzyme activities, measured in an *in vitro* test system, are strongly inhibited in the presence of ion concentrations exceeding 100 mM. Yet, salt-tolerant higher plants (halophytes) and marine algae contain within their cytoplasm ions (mainly K^+, but also Na^+ and Cl^-) in the range of 350–800 mM, and the concentration of ions in the vacuolar sap is usually higher. The question arises as to how enzymes in the cytoplasm are able to cope with those high-ion concentrations. Enzymes isolated from halophilic prokaryotes (e.g., *Halobacterium*) not only tolerate but need high KCl or NaCl concentrations (up to 1–4 M!) for optimal activity (Brown 1976).

All algal cells investigated accumulate at least one low molecular weight organic compound when subjected to increasing salinities (see Experiment

210

25; Kirst and Bisson 1979). Such compounds (especially polyols and the amino acid proline) not only act as osmotica but also as "compatible solutes" (Borowitzka and Brown 1974), i.e., in high concentrations they do not inhibit cellular metabolism and enzyme activity as ions do. The chemical mechanism of this protective nature is not yet fully understood. Compatible compounds possibly maintain an "artificial watersphere" by interaction with the surface of enzyme proteins.

In this experiment, the property of mannitol as a compatible solute is demonstrated. The activity of malate dehydrogenase (MDH) in an *in vitro* test system is compared in the presence of mannitol or an iso-osmotic concentration of NaCl. MDH was chosen because of its wide abundance in all organisms. Highly active enzyme can be easily extracted from algae. MDH catalyzes the following reaction in the tricarboxylic acid (TCA) cycle:

$$\text{Oxaloacetate} + \text{NADH} + \text{H}^+ \stackrel{\text{MDH}}{\rightleftharpoons} \text{malate} + \text{NAD}^+$$

The reaction is measured with a spectrophotometer, by recording the disappearance of the NADH-absorption peak at 340 nm.

Procedure

Preparing a crude enzyme extract. Micro- and macroalgae: freeze algae with liquid nitrogen. Put 5–10 g frozen algal material into a precooled mortar, add 5–10 mL extraction buffer and some quartz sand, homogenize. Centrifuge with a clinical (desk-top) centrifuge (1000–1200 × g, 5 min) or filter the homogenate, save the supernatant and extract the residue again with 5–10 mL extraction buffer. Centrifuge or filter and combine extracts.

The extract can be stored at −20°C. This extract may be used directly as "crude enzyme extract" in the enzyme assay, or may be partially purified first by ammonium sulfate precipitation.

Enrichment of enzyme by ammonium sulfate precipitation. With this step, a large portion of inactive storage protein and chlorophyll-protein complexes will be removed, and MDH among other enzymes is enriched in the 35%–70% saturated ammonium sulfate fraction. The flow sheet is as follows (all steps at 4°–10°C in ice bath):

Thaw homogenate (crude extract) and centrifuge (refrigerated centrifuge; 3000–5000 × g; 5 min).

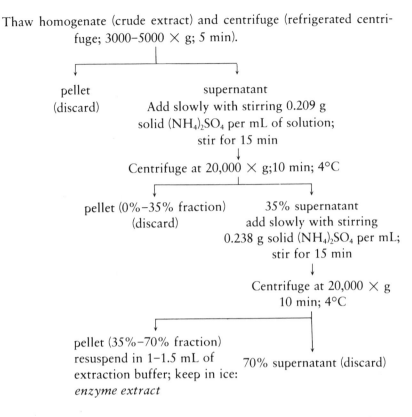

pellet
(discard)

supernatant
Add slowly with stirring 0.209 g
solid $(NH_4)_2SO_4$ per mL of solution;
stir for 15 min

Centrifuge at 20,000 × g;10 min; 4°C

pellet (0%–35% fraction)
(discard)

35% supernatant
add slowly with stirring
0.238 g solid $(NH_4)_2SO_4$ per mL;
stir for 15 min

Centrifuge at 20,000 × g
10 min; 4°C

pellet (35%–70% fraction)
resuspend in 1–1.5 mL of
extraction buffer; keep in ice:
enzyme extract

70% supernatant (discard)

Enzyme assay. Depending on the algal material used, a dilution of the
35%–70% fraction (up to 1:150) might be necessary; hence a "pre-test" is
recommended. If the reaction is complete after less than one minute,
enzyme concentration is too high; dilute as appropriate with extraction
buffer.

1. Assay in duplicate to estimate full enzyme activity (100%): Volumes
given are for microcuvettes and yield 1 mL total volume. If normal-size
cuvettes are used, the total volume will be 3 mL and the amounts of each
solution must be tripled:

Sample cuvette: 100 μL assay buffer
30 μL NADH stock
5 μL enzyme extract (crude or enriched)
765 μL H_2O

The final concentration in the assay is: Buffer 100 mM, NADH 0.15 mM.
Blank cuvette: 1 mL distilled water or (if crude extract is used) the assay
mixture minus oxaloacetate.

Read extinction (absorbance) in spectrophotometer at 340 nm. Start the reaction by adding 100 μL oxaloacetate stock to the sample cuvette; t = 0; mix well, final concentration 0.25 mM. Usually, reaction is completed after one to four minutes depending on the algal species and the efficiency of extraction procedures. According to the result of the "pre-test," a fixed reaction time (two or three minutes; reaction *need not* be completed) must be used for all subsequent tests. After this time, read decrease of extinction (absorbance).

Calculate average of activity. Enzyme activity (E) is (for simplicity) expressed as:

$$E \cdot min^{-1} = \frac{A_o - A_t}{t}$$

(A_o: Absorbance before start; A_t: Absorbance after time t; t: time in minutes).

2. Assays in presence of NaCl or mannitol (two different concentrations each):

Effect of NaCl on enzyme activity: Replace H_2O in the assay described above as follows:

    a) 250 μL NaCl stock $\Big\}$ yielding final NaCl concentration
       515 μL H_2O of 125 mM in the test

and

    b) 500 μL NaCl stock $\Big\}$ final NaCl concentration 250 mM
       265 μL H_2O

Effect of mannitol on enzyme activity: Replace H_2O in the assay described above as follows:

    a) 250 μL mannitol stock $\Big\}$ yielding final mannitol concentration
       515 μL H_2O of 200 mM in the test

and

    b) 500 μL mannitol stock $\Big\}$ final mannitol concentration 400 mM
       265 μL H_2O

The assays containing 125 mM (250 mM) NaCl or 200 mM (400 mM) mannitol are iso-osmotic. Hence, with respect to osmotic stress on the enzyme there is no difference.

Make three measurements of each concentration, calculate average.

Present data as a table. Express loss of activity in presence of NaCl or mannitol as percent of control enzyme activity (=100%).

Questions

1. Explain the reversible precipitation of proteins by ammonium sulfate.
2. Summarize the main adverse effects of high ion concentrations on enzyme activity.
3. The term "compatible solute" has to be used with care, taking into account that it is based on findings with *in vitro* systems. Considering that (a) many data on ion contents of algal cells are obtained from the whole cell and (b) high ion concentrations inhibit *in vitro* enzyme activities, where might you expect the ions and the organic solutes to be compartmentalized within the cell?

REFERENCES

Borowtizka, L. J. & A. D. Brown (1974). The salt relations of marine and halophilic species of the unicellular green alga, *Dunaliella*. The role of glycerol as a comaptible solute. *Arch. Microbiol.* 96, 37–52.

Brown, A. D. (1976). Microbial Water Stress. *Bacteriological Reviews* 40, 803–846.

Kirst, G. O. & M. A. Bisson (1979). Regulation of turgor pressure in marine algae: ions and low-molecular-weight organic compounds. *Aust. J. Plant Physiol.* 6, 539–556.

Lange, O. L., P. S. Nobel, C. B. Osmond, Z. Ziegler (1983). Physiological Plant Ecology III: Responses to the Chemical and Biological Environment. *Encyclopedia of Plant Physiology, New Series.* Springer-Verlag, Berlin, Heidelberg, New York.

NOTES FOR INSTRUCTORS

Equipment

1. centrifuge for large quantities (4 × 250 mL bottles) (1), for harvesting cultures of microalgae
2. clinical (desk top) centrifuge (1)
3. refrigerated high-speed centrifuge (20,000 × g) (1)
4. thermos flask for liquid nitrogen (1)
5. balance (1)
6. ice bath (1)
7. spectrophotometer to read UV (340 nm) (1)
8. magnetic stirrer (1)
9. rack for centrifuge tubes (1)
10. stopwatch (1)

Glassware

1. Mortar and pestle (1)
2. centrifuge tubes for desk-top centrifuge (8)
3. centrifuge tubes for high speed (20,000 $\times$ g) (4)
4. measuring cylinders: 25 mL, 50 mL and 100 mL (1 each)
5. pipettes: 10 mL and 5 mL (2)
6. pipettes 2 mL (3)
7. pipettes 1 mL ("enzyme pipettes") (3)
8. micro pipette(s), variable 5 μL to 200 μL (1 or 2)
9. beakers 50 mL for protein precipitation (2)
10. Quartz-cuvettes; d = 1 cm (microcuvettes 1.5 mL or normal size 4.5 mL) (4)
11. plastic rods for mixing the assay in the cuvettes (2)

Chemicals

All chemicals of analytical grade; e.g., Sigma grade I, Boehringer-Mann-heim grade I.

1. Buffers and stock solutions needed:
   a) 200 mL extraction buffer: 50 mM Tris-HCl, pH 7.5
   b) 10 mL assay buffer: 1 M Tris-HCl, pH, 7.5, plus 33 mM MgSO$_4$
   c) 2 mL oxaloacetate: 2.4 mM (prepare fresh daily)
   d) 2 mL NADH: 5 mM
   e) 5 mL mannitol: 0.8 M
   f) 5 mL NaCl: 0.5 M
2. Quartz sand
3. ice
4. liquid nitrogen
5. artificial sea salt for culture media

Algal material

Microalgae. (for culture methods see Experiment 25). Harvest cultures of unicellular algae, such as *Platymonas* sp., *Dunaliella* sp., *Chlorella* sp., etc. by centrifugation and combine the pellets. Two- to 5 L of culture with about 2–4 $\times$ 10^6 cells mL^{-1} will give 3–10 g fresh weight. Resuspend the pellet once with 50 mL extraction buffer and centrifuge. Freeze the algal pellet with liquid nitrogen and store at -20°C, if necessary. If a French press is available, homogenates of the microalga can be obtained as follows: To a known amount of frozen algal cells, add double the amount of extraction buffer. Passage through the French press should be slow, and the eluate caught in a flask sitting in ice. After centrifugation, the supernatant may be used as crude extract or partially purified with $(NH_4)_2SO_4$, as mentioned previously.

Macroalgae. All green algae are suitable, e.g., *Ulva* sp., *Enteromorpha* sp., *Chaetomorpha* sp.; Phaeophyceae and Rhodophyceae are less suitable because of the mucilage. Clean the algae carefully to remove epiphytes and cut tissue in small pieces. If necessary a short wash in 80% methanol might remove a great part of any mucilage. Then freeze with liquid nitrogen and store at $-20°C$.

Because of the high variability of the algal material (especially the mucilage-containing kelps), no precise recipes can be given for the assay. The amount of algae needed and the dilution of the enzyme extract must be worked out for each species in a pre-test by the instructor. Freezing with liquid nitrogen is not absolutely necessary but is useful in cracking of the cell walls to improve extraction.

Comments

Other enzyme systems are also suitable for this experiment and will be present in the 35%–70% fraction, e.g., glucose-6-P-dehydrogenase, hexokinase, etc. The ammonium sulfate precipitation is time consuming and hence might be omitted. The crude extract from the homogenate can be used; in this case the blank in the spectrophotometer must contain as much enzyme extract as the assay cuvette because of absorbance. All solutions needed in the enzyme assay should be stored in a refrigerator.

Using the MDH-system isolated from *Platymonas,* we found a loss of activity in presence of NaCl of ca. 30% (125 mM) and more than 50% (250 mM), but with mannitol only 5%–10% (200 mM) and 15% (400 mM).

Scheduling

Full procedure runs 3–4 h; using crude extract, one to one and a half hours.

25

Accumulation of organic compounds with increasing salinity: mannitol accumulation in the unicellular phytoplankter *Platymonas*

G. O. *Kirst*

Fachbereich 2 (Biologie-Chemie) – Botanik, University of
Bremen, D-2800 Bremen (FRG)

FOR THE STUDENT

Introduction

Marine algae subjected to changes in salinity regulate their internal concentrations of the main ions (Na^+, K^+, Cl^-) and additionally the content
of low-molecular-weight organic compounds so as to maintain a constant
turgor pressure (see Experiment 23). Usually the main photosynthetic
product is involved, e.g., sucrose or polyols (mannitol, glycerol) in Prasinophyceae and Chlorophyceae, mannitol in Phaeophyceae, and floridoside
or digeneaside in Rhodophyceae. With hyperosmotic stresses (increasing
salinity), those compounds are accumulated while under hypo-osmotic
conditons (decreasing salinity)–especially after sudden dilution of the
media ("down shock")–the organic solutes might be released into the
medium. In the dark, storage products (large osmotically inactive molecules) are hydrolyzed to low-molecular-weight compounds when the alga
is subjected to increasing salinity (Reviews: Hellebust 1976; Kauss 1977).
Platymonas (Tetraselmis) sp. is used as a test organism here. It is abundant
in the coastal phytoplankton and is widely cultivated as a food source for

fish and oyster breeding. The organism is very salt tolerant (see Kirst 1977) and accumulates mannitol under hyperosmotic stress.

In this experiment, the time course of mannitol accumulation in *Platymonas* is investigated after a sudden increase of the NaCl concentration in the medium. Mannitol is estimated in the extracts according to the periodate oxidation method of Lewis and Smith (1967):

$$
\begin{array}{c}
\text{H} \\
\text{HC} - \text{OH} \\
| \\
\text{HO} - \text{CH} \\
| \\
\text{R}
\end{array}
\quad + \text{HIO}_4 \rightarrow
\begin{array}{c}
\text{H} \\
\text{HC} = \text{O} \\
\end{array}
+ \text{O} =
\begin{array}{c}
\text{CH} \\
| \\
\text{R}
\end{array}
+ \text{HIO}_3 + \text{H}_2\text{O}
$$

In acid solutions (pH 4–5), mannitol is oxidized completely within the first minute following addition of periodate, while other polyols and sugars are not affected so quickly. Periodate has a distinctive absorption peak at 260 nm (UV-light). Hence, its decrease during mannitol oxidation can be followed directly by a spectrophotometer. The disappearance of periodate is proportional to the amount of mannitol present.

Procedure

Hyperosmotic stress. Combine into a labeled 250 mL Erlenmeyer flask 35 mL of the algal suspension (stock suspension: *Platymonas* cultures harvested by the instructor) with an equal amount of stock solution consisting of seawater medium plus 1 *M* NaCl. This results in a final concentration of 0.5 *M* NaCl above seawater (approximately double-strength seawater). Take two 5 mL samples right away (time t = 0). Transfer each sample into a separate and labeled centrifuge tube and proceed immediately with mannitol extraction (see following). Set the Erlenmeyer flask with the suspension in the light. The algal suspension in the flask must be bubbled with air or shaken on a shaker throughout the experiment. Withdraw two 5 mL samples at each time: t = 15, 30, 45, 60 min. The exact time should be noted.

Set up a control culture in similar manner as follows: At t = 2 min, mix 11 mL of the algal stock suspension with 11 mL of normal seawater medium in a labeled 50 mL Erlenmeyer flask. This results in a dilution of the suspension of the stressed algae, but the salinity remains unchanged. Treat the control as the stressed algae: set in light and bubble suspension with air or shake. Collect two 5 mL control samples at t = 32 min and at the end of the experiment: t = 62 min. The exact times should be noted.

Extraction of mannitol. Spin down the 5 mL samples at 1200 × g (top speed of a clinical lab centrifuge), 2 min, immediately after collection. Discard supernatant and resuspend the pellet in 2.5 mL of 40% methanol. Extract cells for 10 min in a waterbath at 70°C. Cover the centrifuge tubes during heating with a glass marble or foil to minimize evaporation. Centrifuge methanol extract at 1200 × g, 2 min, and save the supernatant for mannitol estimation in labeled vials or test tubes.

Mannitol estimation: Prepare a calibration curve according to the following table pipetting in order from left to right:

H₂O (mL)	Mannitol standard (mL)	Acetate buffer (mL)	Start: Periodate (mL)
1	0	1	1 (periodate blank)
0.9	0.1	1	1
0.8	0.2	1	1
0.5	0.5	1	1
0	1.0	1	1

Pipette directly into quartz-cuvettes (d = 1 cm); start reaction by adding the periodate solution (t = 0), mix well and read sample after exactly 60 seconds at 260 nm in a spectrophotometer. Use a mixture of 2 mL H₂O and 1 mL acetate buffer as reference. Leave enough time between each start to read each at end. Draw the graph of Absorbance versus µmol of mannitol, where Absorbance is calculated as difference between absorbance of the periodate sample and absorbance of the mannitol standard.

Measure the algal extracts by pipetting directly into quartz-cuvettes:

	H₂O (mL)	Sample (algal extract) (mL)	Acetate buffer (mL)	Start: Periodate (mL)
Periodate blank	1	—	1	1
Sample blank	1	1	1	—
Sample	—	1	1	1

Follow same procedure as with the calibration curve: Start by adding the periodate solution, mix well, read after 60 seconds. Use a mixture of 2 mL H_2O and 1 mL acetate buffer as reference. A blank of the sample is needed to correct for its inherent absorbance at 260 nm.

Calculate the absorbance ΔS as: $\Delta S = (Bp + Bs) - S$, where Bp is the absorbance of the periodate blank, Bs the absorbance of the sample blank, and S the absorbance of the sample. Then use ΔS to read the mannitol content from the calibration graph. The quantitative estimation of mannitol is not a linear relation with high concentrations. If the range of the calibration standards is exceeded, the sample must be diluted.

Calculations

Calculate for each sample the total amount of mannitol of the 5 mL algal suspension (remember that there were 2.5 mL of methanol extract).

Present data as a graph of μmol mannitol per 5 mL sample versus time. For each time, give the values of both samples to show the degree of variation.

Questions

1. Mannitol is synthesized during photosynthesis in *Platymonas*. Fructose-6-phosphate is a precursor and mannitol is interconvertible with starch. Design a simple scheme showing the possible sites of control of the mannitol pool by enzyme activation or inhibition during hyperosmotic and hypo-osmotic stress.
   What do you expect from the mannitol pool when photosynthesis is inhibited (e.g., by DCMU) and the algae are stressed hyperosmotically?
2. What chemicals are polyols, mono-, di-, and oligosaccharides, heterosides, polysaccharides? Give the structural formula for one example of each.

REFERENCES

Hellebust, J. A. (1976). Osmoregulation. *Annual Reviews of Plant Physiology*, 27, 485–505.

Kauss, H. (1977). Biochemistry of osmotic regulation. In, S. H. Northcote (ed.) *International Review of Biochemistry–Plant Biochemistry II* Vol. 13. University Park Press, Baltimore.

Kirst, G. O. (1977). Coordination of ionic relations and mannitol concentrations in the euryhaline alga *Platymonas subcordiformis* after osmotic shocks. *Planta*, 135, 69–75.

Lewis, D. H. & D. C. Smith (1967). Sugar alcohols (polyols) in fungi and green plants. II. Methods of detection and quantitative estimation in plant extracts. *New Phytol.* 66, 185–204

NOTES FOR INSTRUCTORS
Equipment

1. centrifuge for harvesting the algae (capacity 4 × 250 mL or more) (1)
2. clinical (desk-top) centrifuge (1)
3. spectrophotometer (double beam; UV; 260 nm required) (1)
4. waterbath (70°C), racks for centrifuge tubes (1)
5. shaker or air pump with tubing (for shaking or bubbling the cultures during the experiment) (1)
6. stopwatch (1)
7. light source

Glassware

1. 1L Erlenmeyer flasks for growing the algae (or cultivating flasks) (5)
2. 250 mL Erlenmeyer flask (1)
3. 50 mL Erlenmeyer flask (1)
4. 1 L volumetric flask (2)
5. 500 mL volumetric flask (1)
6. 200 mL volumetric flask (1)
7. 100 mL measuring cylinder (1)
8. 25 mL measuring cylinder (1)
9. 250 mL centrifuge bottles for harvesting the algae (4)
10. centrifuge tubes (10 mL) (16)
11. 5 mL pipettes (2)
12. 1 mL pipettes (5)
13. 0.5 mL pipettes (reading 0.1–0.5 mL) (2)
14. glass marbles or foil
15. set of 4 Quartz-cuvettes (1)

Chemicals

1. natural filtered seawater or artificial seawater (e.g., ready-mix used for fish tanks)
2. NaCl
3. Na-metaperiodate
4. acetic acid (conc.)
5. NaOH solution 2 N

6. mannitol (analytical grade)
7. methanol 40% (100 mL)

Solutions

1. Acetate buffer 1 M, pH 4.5: 12 mL of acetic acid (conc.) is added to 100 mL distilled water and brought to pH 4.5 with 2 N NaOH. The final volume is made to 200 mL with distilled water. Store in refrigerator.
2. Na-periodate, 3.5 mM: Dissolve 375 mg Na-metaperiodate in 500 mL quartz distilled water. Store in refrigerator.
3. Mannitol standard: Dissolve 91.1 mg mannitol in distilled water and make up to 1 L, yielding a 0.5 mM stock solution. Store in refrigerator.
4. Stock solution for hyperosmotic stress: 1 mol NaCl (58.4 g) is dissolved in seawater and made up to 1 L with seawater.

Algae

Grow *Platymonas* sp. from stock cultures in filtered or artificial seawater (30‰–35‰). The cultures should be bubbled with air or stirred and kept under continuous light (ca. 150–300 μE m^{-2} s^{-1}). Dense olive-green cultures are obtained after five to seven days. Harvest by centrifugation (100–150 $\times$ g for 5 min) at least 3 h or up to one day prior to experiment. Resuspend the algal pellets in a little fresh medium and combine them. Bring the culture to a density of 2–4 $\times$ 10^7 cells mL^{-1} by appropriate dilution with medium. A total amount of 50 mL should be prepared for the experiments and kept vigorously bubbled with air or stirred and illuminated until use.

Cultivation of *Platymonas* from a purchased stock culture maintained on agar: If the cells are inoculated in large amounts of medium they usually don't grow; a precultivation in 10–20 mL medium is recommended until a dense green suspension is obtained, which is transferred to larger volumes. Vigorous bubbling or stirring from time to time prevents clumping of the cells, which occurs with rising pH (above 9.5). Though the buffer capacity of natural seawater is usually sufficient for the duration of culture and experiments used here, a check of pH and resetting to 8.2–8.4 by bubbling with CO_2 or air enriched with CO_2 might be necessary, especially in dense cultures. Artificial seawater should be enriched in any case with NaHCO$_3$ (1 mM final concentration).

The culture density of 1 − 2 $\times$ 10^7 cells mL^{-1} in the experiment is optimal; however, suspensions of 10^6 cells mL^{-1} will also work. If material is in short supply, draw fewer samples rather than dilute the suspension to less than 0.5×10^6 cells mL^{-1}.

Comments

Modification of the experiment: Estimation of mannitol content of cultures of *Platymonas* sp. grown at various salinities. The mannitol content has to be related to a common basis: e.g., cell number or packed cell volume because of the variation of growth in the different cultures.

The method of mannitol estimation can also be used with kelps. Extraction of the tissues with 80%–90% methanol is needed to minimize the interference of the mucilage. However, responses of kelp tissues to increased salinities are very slow and long incubation periods (four days) are necessary. Basically, the osmotic shock experiments described above can be done with other unicellular euryhaline algae, e.g., *Dunaliella*. However, the organic solute is different and in the case of *Dunaliella* glycerol has to be estimated. This is better done enzymatically using a commercial kit (supplied by Boehringer-Mannheim or Sigma).

Scheduling

Platymonas should be cultivated one week before the experiments start. The experiment fits in a 3 h lab period. If necessary, the extracts (well sealed) can be stored in a freezer for several weeks.

26

The contractile vacuole as an osmoregulatory organelle

G. O. Kirst

Fachbereich 2 (Biologie-Chemie) - Botanik, University of Bremen, D-2800 Bremen (FRG)

FOR THE STUDENT

Introduction

In freshwater environments, the osmotic gradient between a cell with a high concentration of internal solutes and the medium will cause an influx of water that would result in swelling and, ultimately, rupture of the cell. In plant cells possessing a proper cell wall, swelling is prevented by the rigidity of the wall, and a turgor pressure develops instead. In cells that lack continuous cell walls, a contractile vacuole complex serves to eliminate fluid, compensating for the osmotic influx of water (Hansmann and Patterson 1984). In wall-less organisms living in media of high osmotic pressure, e.g., sea water, contractile vacuoles are not needed. Many freshwater algal flagellates, including volvocales, euglenoids, chloromonadophytes, and cryptophytes, typically contain one or two contractile vacuoles. The vacuoles discharge their contents at a fixed point in the surface of the cell. The contractile vacuole works in two phases: filling (diastole) is followed by expulsion (contraction: systole). During filling, many small accessory vesicles fuse to a large central vacuole, which finally rounds up. The mechanism of "water pumping" is explained by an active ion (or solute) transport, located in the membranes of the vesicle. This in turn draws water across the membrane into the vesicle. The filled central vacuole is "squeezed out" by contractile filaments, closely associated with the vacuole, which can be seen on electron micrographs. During the systole the vacuole fuses into the plasma membrane and disappears from view under the microscope. From this moment on, time can easily be measured until the next systole occurs. The time to complete one cycle varies. It depends upon nutritional state, temperature and species. A cycle may take 10–30 sec (*Chilomonas* sp., *Euglena* sp.) or last 50–90 sec. However, under given conditions the pulsing frequency is fairly constant.

In this experiment the contraction cycle in algal cells under standard growth condition is first estimated. The cells are then subjected to increasing salinity. This will increase the time measured for one cycle, i.e., the pulsing frequency will decrease. At a certain salinity the contractile vacuole will cease to function.

Procedure

1. Osmotic stress: Mix a dense algal suspension with NaCl or sucrose stock solution in the following ratios (volume/volume) to yield a series of increasing osmolality:

Number	1	2	3	4	5	6	7	0 (control sample)
Culture (mL)	9.5	9	8.5	8	7	5	3	10
Stock NaCl or sucrose (mL)	0.5	1	1.5	2	3	5	7	0

2. Microscopic observations: Place one drop of a sample of the concentration series on a microscopic slide and slow the algae down using the material your instructor has provided. (Take into account that this drop of solution dilutes the sample by about 1:1.) Cover with cover slip and seal edge of cover slip with a ring of paraffin oil to prevent the solution from evaporating. Observe at high magnification (40 × objective) after appropriate orientation at low magnification. Quickly check through the series starting with the control sample to find the range of concentration in which a contractile vacuole is still observable.

Make a rough estimate of the external osmotic pressure at which the contractile vacuole stops functioning. To a good approximation, osmotic pressure from freshwater equals 0 MPa and rises linearly to 0.92 MPa, which is the osmotic pressure for the NaCl or sucrose stock solution (e.g., No. 6 of the preceding dilution series will exhibit an osmotic pressure of about 0.46 MPa). (For definition of MPa (Megapascal) see Experiment 23.)

3. Pulsing frequency: Select in addition to the control sample another slide with algae under osmotic stress but with a still–working contractile vacuole.

Measure the duration of a complete cycle (from systole to systole) in seconds. Calculate the average of three measurements and repeat this with three to five (or more) organisms of each sample.

Give data as mean $\pm$ SD (n) (n = number of organisms) and compare results

obtained with normal and stressed algae.

Questions

1. The osmotic stress is achieved in the experiment using two different types of solutes, but in iso-osmotic concentrations. Explain the basic difference of the solutes and the possible effects on the organisms.
2. Mutants of *Chlamydomonas moewusii* lacking a contractile vacuole exist. Not surprisingly, this is a freshwater species. Design a suitable medium for these mutants.

REFERENCES

Hausmann, K. & D. J. Patterson. (1984). Contractile vacuole complexes in algae. In, W. Wiesner, D. Robinson & R. C. Starr (eds) *Compartments in Algal Cells and Their Interaction*. Springer-Verlag, Berlin, pp. 139–146.

Heyward, P. (1978). Osmoregulation in the alga *Vacuolaria virescens*. Structure of the contractile vacuole and the nature of its association with the Golgi apparatus. *J. Cell Sci*. 31, 213–24.

NOTES FOR INSTRUCTORS
Equipment

1. microscope (1)
2. stopwatch (1)
3. test tube rack (1)

Glassware

1. microscope slides (1)
2. cover slips (2)
3. test tubes or Erlenmeyer flasks (20–50 mL) for the concentration series (10)
4. glass rod or brush (for paraffin oil ring) (1)
5. 10 mL pipettes (2)
6. 5 mL pipettes (2)
7. 1 mL pipettes (2)

Chemicals

1. NaCl, 200 m*M*
2. sucrose, 350 m*M* (about iso-osmotic with the NaCl solution)
3. gelatine or agar or carboxymethyl cellulose
4. paraffin oil

Algae

All freshwater unicellular algae with good visible contractile vacuoles are suitable. Most of them are flagellates and as such highly mobile. To facilitate the observations, the algal cells must be trapped by putting some cotton wool into the drop with the algal suspension. The cells will be trapped between the threads. Another method to slow the algae down is to increase the viscosity of the medium by adding a drop of a stock solution of 3% gelatine or carboxymethyl cellulose (alternatively: 0.5%–0.7% agar or "quince slime" obtained by soaking a few quince seeds in a little water). These substances are osmotically inactive.

Euglenoids provide good cells because the contractile vacuole is situated close to the eye-spot (stigma). However, many *Euglena* sp. undergo "euglenoid movement" or metaboly (changing cell shape) when not swimming, which again might cause difficulties during observation. *Euglena acus* and *Phacus* spp. exhibit little or no such movement. Use dense cultures to increase the probability of finding an organism in a fixed position and containing an easily observable contractile vacuole. A little patience on the part of the students will be rewarded.

Comments

The students should be supplied with dense cultures of flagellates. Enrichment of motile algae is best done with a light-trap: Cover the culture flask with black foil except the top 2–3 cm of the medium and illuminate from the side. Within a few hours the algae will collect in the illuminated zone of the medium and can be harvested by pipetting. Centrifugation of the culture is not suitable unless the algae are allowed to recover for 2–3 h.

Instead of using an alga with a chloroplast, which might hide the vacuole, plastid-less protists, such as *Paramecium* sp., are even better. Chloroplast free cells of *Euglena* sp. may also work but these organisms have not as yet been tested.

Scheduling

1.5–2 h laboratory work.

Relative gas composition of the air bladders (pneumatocysts) of kelps and fucoids

G. O. *Kirst*

Fachbereich 2 (Biologie-Chemie) - Botanik, University of
Bremen, D-2800 Bremen (FRG)

FOR THE STUDENT

Introduction

The thalli of several brown seaweeds develop air bladders (vesicles or pneumatocysts), which serve as floats to maintain the distal portion of the alga buoyant in the water volume. The air bladders are distributed in pairs within the flat fronds of *Fucus vesiculosus* or found as periodic swellings of the axes, as in *Ascophyllum nodosum*. The gas composition of the bladders varies and consists mainly of nitrogen, oxygen and carbon dioxide. The stipe of *Nereocystis* (several meters long) is terminated by a huge spherical pneumatocyst, which is up to about 15 cm in diameter. This bladder contains a mixture of gases, carbon monoxide being one of the components (Foreman 1976).

The method used here to estimate the relative gas composition of the bladders is a modification of a micro gas-absorption method applied to measure the gas composition of the aerenchyma (lacunar system) in rhizomes and shoots of various aquatic angiosperms (Haldemann and Brändle 1983). The method is based on the absorption of gaseous CO_2 in strong alkaline solution and of oxygen in pyrogallol solution. The change of the size of a gas bubble trapped in a microcapillary (pipette) is estimated under the microscope after changing the solution surrounding the bubble in the capillary.

228

Procedure

Collection of gas samples. Species with air bladders are needed, such as *Fucus vesiculosus, Ascophyllum nodosum, Sargassum* spp., *Macrocystis* spp., *Nereocystis luetkeana.* Place the pneumatocysts under water in a water tank. The pneumatocysts very frequently will have a gas pressure less than atmospheric and would pull in air from the atmosphere upon puncturing. In this way only water will enter. It is essential that there be no dead air space in the gas-withdrawal syringe. This is achieved by having such air space (e.g., in the needle) filled with the citric acid solution: Draw some into the syringe and then expel with all air bubbles. The gas samples are obtained directly by withdrawing about 50 μL of gas from the bladder with a gas-tight syringe.

Alternatively, larger amounts of gas can be caught if the tissues are cut open under a funnel submersed in a water-filled tank, which leads the gas bubbles into a water-filled graduated test-tube or vial (Fig. 27.1a). No gas must be trapped in the system before the start. The test-tube (or vial) is then sealed under water with a serum cap trapping the gas and the remaining water. From this sample, suitable amounts of gas can be withdrawn through the serum cap with a syringe. Store test tube upside down when

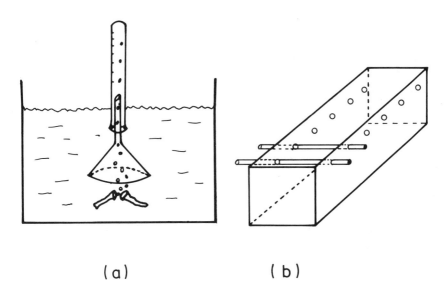

(a)   **(b)**

Fig. 27.1. (a) Set-up of a system to trap gas bubbles squeezed out of tissues cut open under water. (b) Plastic box with holes to hold micropipettes.

not in use. The water in the water tank and test-tube should be slightly acidified (pH 2-4) with a few drops of sulfuric acid to avoid absorption of CO_2 possibly present in the samples.

Measurement of relative gas composition. Place calibrated micro-pipettes (micro caps) into labeled opposite holes of suitable plastic boxes (Fig. 27.1b). A set of 10 to 15 of such pipettes will do. The size of the pipettes depends on the amount of gas available. Fifty μL or 100 μL pipettes are usually sufficient. For smaller samples, 10 μL or 20 μL micro caps are recommended.

Fill pipettes with a solution of 1 M citric acid. The solution will be sucked into the micro caps out of a larger pipette (1 mL) by capillary forces. Fill plastic box with water at room temperature. This will serve as a buffer against small temperature fluctuation.

Now inject with a syringe a gas bubble of about 10 to 30 μL into the micro cap (2 to 5 μL are used for the small pipettes). Repeat until all but one or two pipettes in the box contain a gas bubble, which may differ in size. Inject into the remaining one or two pipettes normal air to be used as a control.

Leave for 2–3 min to equilibrate. Measure the length of each bubble using a stereomicroscope or microscope with a micrometer eyepiece. Calibration of the micrometer is not necessary, since the length is read relative to the starting length.

Replace the citric acid solution, which enclosed the bubble, by a solution of 4 M KOH. Care must be taken that the gas bubble does not escape from the capillary. The exchange of the solutions is best done as follows: Remove from one end of the capillary a small portion of the citric acid with a syringe and replace the withdrawn liquid by the KOH-solution. Do the same from the other end. Repeat several times. While solutions are withdrawn and refilled the gas bubble will move a bit in the capillary and help to mix the KOH with the remaining citric acid solution. The strong alkali neutralizes the citric acid and further increases the pH.

Allow for a short period of equilibration of 2–5 min, during which the box is carefully tilted so that the gas bubbles move back and forth a bit in the pipettes. Repeat tilting procedure until there is no change in gas bubble volume.

Measure again the length of each bubble. Typically a small but distinct decrease is observed with the gas samples from the plants, but almost none with the air bubble used as control (why do these samples show very little or no decrease?). Express the difference of the lengths of each sample as percent of the original size (= 100%).

Replace the 4 M KOH solution by a mixture of 2 M KOH and 5% pyrogallol. Repeat procedure as described above. Measure lengths of the bubbles and calculate percent of the differences in relation to the original size. This will represent the percentage of O_2 in the gas mixture. The remaining length of each bubble is also expressed as percent of the original length and is a measure of nitrogen and other gases.

Present the data in a table (mean ± SD [n = number of samples]). Compare control data (air bubble) with the theoretical composition of air: CO_2 0.03%, O_2 21%, N_2 78%; and evaluate the reliability of the method.

Questions

1. How will the gas composition change in air bladders or aerenchymas of the following tissues after a long period of (a) light or (b) dark exposure: algal frond, leaf of seagrasses, rhizome of seagrasses?
2. Algae possessing gas bladders are abundant in the upper littoral region or, if they start in deep waters, their fronds are close to the surface of the water. Explain the absence of gas bladders in deep waters.

REFERENCES

Haldemann, C. & R. Brändle (1983). Avoidance of oxygen deficit stress and release of oxygen by staled rhizomes of *Schoenoplectus lacustris. Physiol. Veg.* 21, 103–113.

Foreman, R. E. (1976). Physiological aspects of carbon monoxide production by the brown alga *Nereocystis luetkeana. Can. J. Bot.* 54, 352–360.

NOTES FOR INSTRUCTORS

Equipment

1. A small plastic box (polyethylene or other *soft* plastic material) prepared as shown in Fig. 27.1b: Holes drilled in opposite sites to hold the micropipettes. The micropipettes should fit tightly into the holes.
2. stereo-microscope with micrometer
3. gas-tight syringes (100 μL; as used for gas chromatography) (2)
4. 1-mL (or less) all-glass syringes with luer needles (size: # 22 or # 24) (2) (disposable plastic syringes tend to accumulate surface-tension air bubbles)
5. serum caps

Glassware

1. micro pipettes (micro caps) of various sizes: 10, 20, 50 and 100 μL
2. glass funnel (1)
3. water tank (1)
4. vials or test-tubes

Chemicals

1. 1 N sulfuric acid to acidify water in the water tank
2. 100 mL 1 M citric acid
3. 50 mL 4 M KOH (stored in plastic bottle)
4. 50 mL 5% (w/v) pyrogallol in 2 M KOH

The pyrogallol solution must be kept under mineral oil (paraffin oil): Place dry pyrogallol in a reagent bottle, cover with mineral oil and add the 2 M KOH through the oil. This keeps exposure to oxygen to a minimum. When the solution becomes very dark brown to black it should be discarded.

In all solutions there must be no/minimal dissolved gas that could be liberated: Bring all solutions to ambient temperature before use, and, even better, heat to about 80°–90°C to drive off gas, or place under vacuum to remove dissolved gas.

Use of 4 M KOH can cause problems. If students do not watch carefully, syringes "seal up" and needle tips block.

Scheduling

2 h.

28

Photoperiodic and blue-light mediated photomorphogenetic effects

Klaus Lüning

Biologische Anstalt Helgoland Zentrale, Notkestrasse 31,
D-2000 Hamburg 52, Federal Republic of Germany

FOR THE STUDENT

Introduction

Light is used by the plant as an energy source for photosynthesis, or as an environmental signal for synchronizing the plant's development with cyclic changes in the environment. The latter aspect is exemplified by the phenomena of photoperiodism (development regulated by daylength) and photomorphogenesis (development regulated by light quality) (Lüning 1981a; Dring and Lüning 1983; Dring 1984).

Plants with a photoperiodic reaction possess an "early warning system"; e.g., deciduous trees form winter buds in short days in autumn, long before the first snow falls. Highly sensitive, nonphotosynthetic sensor pigments are used to detect light as an environmental signal, e.g., the predominantly red- /far-red-sensitive pigment phytochrome mainly in higher plants, or the blue-light sensitive cryptochrome mainly in lower plants (see reviews in Shropshire and Mohr 1983; papers in Senger 1984). The chemical nature of cryptochrome is unknown, but it is probably a flavin.

Photoperiodism in algae. The effects of daylength are particularly obvious in cases where short days induce a qualitative progress in development, such as the transformation from the vegetative to the reproductive stage, the formation of upright thalli from a simple crustose thallus, or the annual formation of new thallus parts, e.g., the new blade in a kelp. Most

233

Table 28.1. *Photoperiodic short-day reactions in seaweeds. Night-break regime effective. R red alga, B brown alga, G green alga, (H) heteromorphic life history in red and green algae, TS tetrasporophyte*

Reaction	Species	Reacting stage	Authors
Formation of conchosporangia	R(H) *Porphyra tenera*	Conchocelis (sporophyte)	Dring (1967a,b); Rentschler (1967)
	R(H) *Bangia atropurpurea*	Conchocelis (sporophyte)	Richardson (1970)
Formation of tetrasporangia	R *Acrochaetium asparagopsis*	isomorphic tetrasporophyte	Abdel-Rahman (1982)
	R(H) *Rhodochorton purpureum*	tetrasporophyte	Dring and West (1983)
	R(H) *Bonnemaisonia hamifera*	Trailliella-phase (TS)	Lüning (1980b, 1981b)
	R(H) *Bonnemaisonia asparagoides*	Hymenoclonium-phase (TS)	Rueness and Asen (1982)
	R(H) *Asparagopsis armata*	Falkenbergia-phase (TS)	Oza (1976); Lüning (1981b)
	R(H) *Calosiphonia vermicularis*	Hymenoclonium-phase (TS)	Mayhoub (1976)
	R *Halymenia latifolia*	isomorphic tetrasporophyte	Maggs and Guiry (1982)
	R(H) *Gigartina stellata*	Petrocelis-phase (TS)	Guiry and West (1983)
	R *Gigartina acicularis*	isomorphic gametophyte	Guiry and Cunningham (1984)
	G(H) *Monostroma grevillei, M. undulatum*	Codiolum-phase	Lüning (1980b)
Formation of gametangia			
Formation of zoospores			
Formation of receptacles	B *Ascophyllum nodosum*	macrothallus	Terry and Moss (1980)
Formation of erect thalli	R *Dumontia contorta*	crustose stage	Rietema and Klein (1981)
	B *Scytosiphon lomentaria*	crustose stage	Dring and Lüning (1975a)
	B *Petalonia fascia, P. zosterifolia*	crustose stage	Lüning (1980b)
Formation of a new blade	R *Constantinea subulifera*	gametophytes and TS	Powell (1964)
	B(H) *Laminaria hyperborea*	sporophyte	Lüning (1986)

of the photoperiodic responses so far discovered among algae are short-day responses, and examples are cited in Table 28.1. You may use this table as a guide to discover more photoperiodic responses among the algal species of your own local flora. Species that seasonally form new outgrowths, or seasonally switch between a macroscopic and a more cryptic generation, are good candidates. If some of the species cited in Table 28.1 are present along your coast, you may test the photoperiodic behavior of your local strains. This may furnish interesting results, because in a particular species there may be considerable genetic variance in regard to photoperiodic behavior of local strains. In the brown alga *Scytosiphon lomentaria,* for example, the critical daylength for erect thallus formation by crusts increases with increasing latitude (Lüning 1980b), a trend well-known from higher plants and animals. However, there are also nonphotoperiodic strains of this alga (Clayton 1978; Kristiansen 1984; tom Dieck 1987).

Photomorphogenetic effects (Fig. 28.1). In brown algae, blue light induces (1) hair formation and two-dimensional growth in *Scytosiphon lomentaria* (Dring and Lüning 1975b) and (2) fertility in the gametophytes of Laminariales (Lüning and Dring 1975; Lüning and Neushul 1978; Lüning

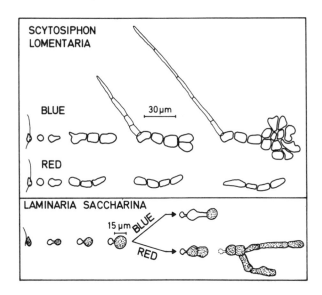

Figure 28.1. Scheme of morphological development of *Scytosiphon lomentaria* and gametophytes of *Laminaria* spp. blue (white) or red light (Dring and Lüning 1983)

1980a; Maier 1984) and Desmarestiales (Müller and Lüthe 1981). (3) In the presence of blue light, kelp female gametophytes do not release eggs during the light cycle of a light-dark regime, but soon after onset of darkness. This effect has been reported for Laminariales in *Laminaria japonica* (Tseng et al. 1959), *Laminaria saccharina* (Lüning 1981c), and *Chorda tomentosa* (Maier 1984), the Desmarestiales in *Desmarestia aculeata* (A. Peters; personal communication), and among the Sporochnales in *Perithalia caudata* (Müller et al. 1985). One would not be surprised if the effect is found to be typical for all members of the three orders.

The action spectrum for the three effects cited above suggests a cryptochrome-like pigment as a sensor pigment, with three absorption peaks or shoulders in the blue, one in the near UV, and no action at wavelengths longer than 520 nm (Fig. 28.2).

Procedures

Photoperiodic effects. The approach is quite straightforward. Using standard techniques, cultivate an alga from spores or zygotes at 8 or 16 h light

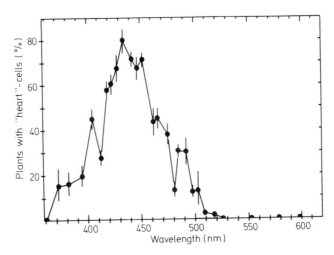

Figure 28.2. *Scytosiphon lomentaria.* Effectiveness spectrum for the induction of two-dimensional growth at 15°C. Plants raised from zoospores and grown in red fluorescent light (20 μE m^{-2} s^{-1}) for 5 d were irradiated for 48 h with 15 μE m^{-2} s^{-1} at each wavelength. Percentage of plants with heart-cells (longitudinally divided apical cells) determined at end of irradiation period. Vertical bars are 95% confidence limits. Additional zero values were recorded at various wavelengths up to 735 nm. (Dring and Lüning 1975b)

per day in two different incubators to look for qualitative differences during the next few weeks. Or, collect adult algal material from the field during the summer and subject samples to short or long days in the laboratory. Since short-day plants require a long night, rather than a short day, it is wise to add to the regimes of 8 or 16 h light per day a night-break regime as a third treatment. Use 8 h light per day plus 1 h light (or even only 1 min!) in the middle of the night and make sure no one opens the culture chamber during the dark cycles. The resulting two short nights will signal "summer" to the alga; consequently, the results you obtain in the night-break regime will be similar to those in the long-day regime. The short-day treatment and the night-break regime are photosynthetically almost equal, and if the algae exhibit dramatically different responses, you may be sure it is a genuine photoperiodic response. Whoever finds a genuine photoperiodic response in unicellular algae will become the discoverer of photoperiodism for this algal group.

Photomorphogenetic effects. Obtain zoospores of *Laminaria* spp., *Desmarestia* spp., and *Scytosiphon lomentaria* as follows:

1. Wipe dry with paper towel a mature, sporangia-bearing thallus piece and store it overnight (only 1–3 h in the case of *Scytosiphon lomentaria*) in a cold room (5°–10°C) in a closed glass vessel, e.g., in a crystallizing dish (9 cm diam, 5 cm high), with no water in it.
2. Immerse thallus piece in Provasoli's enriched seawater (PES) (Starr 1978). Zoospores should be released in few minutes.
3. Pipette zoospores onto coverslips (18 × 18 mm) in Petri dishes at a density of a few hundred spores per coverslip.
4. Cover 1 cm high with seawater medium containing germanium dioxide (4 mL of a saturated solution of GeO_2 in distilled water, added to 1 L of PES) to eliminate diatoms.
5. Change the medium after two days to ordinary PES.
6. Cultivate developing spores on coverslips
   (a) in continuous white light and broad-band red or blue light, at identical photon flux rates in the range of 10–20 μE m^{-2} s^{-1} (as measured with a quanta meter), preferably at 10°C,
   (b) (only *Laminaria* spp., *Desmarestia* spp.:) in a white-fluorescent day/night-regime, e.g., at 16 h light per day.
7. General observations: In blue and white light: formation of hairs, and two-dimensional crusts in *Scytosiphon lomentaria* (during the first week), formation of oogonia and antheridia in *Laminaria* spp. and *Desmarestia* spp. (during the second week). In red light: filamentous thalli, no hairs in *Scytosiphon lomentaria,* no reproductive organs in *Laminaria* spp. and *Desmarestia* spp.

8. Observation of egg release: The day after eggs are first released by gametophytes of *Laminaria* spp. and *Desmarestia* spp., you may observe in the microscope the process of egg release (within 30 sec!) near the end of the next light cycle. Place a green or red interference filter between light source and condensor of your microscope. You should screen the microscope from roomlight, or sit in a dark room and shield the microscope. Egg release starts about 10 min after the onset of "darkness" for the alga (= green or red light), and reaches a maximum after 30–60 min (for details see Lüning 1981c).

Questions

1. What is the ecological advantage of having a photoperiodic reaction (a) in algae with a heteromorphic life history or (b) in a kelp with annual formation of a new blade (see Table 28.1 for examples)? Where do you see indications of an "early warning system"?
2. What is the ecological advantage of the eggs of laminarian gametophytes being released at the onset of darkness? (See Exp. 29 on hormonal systems in Laminariales and Desmarestiales)
3. Several important enzymes with a flavin component, such as nitrate reductase, are activated by blue light (e.g., Azuara and Aparicio 1983; and various papers in Senger 1984). Algal hairs were thought to function as nutrient antennae. One of the ions of major importance for algal growth is nitrate. Hairs are absent in red light in *Scytosiphon lomentaria*. Do you see a possibility to combine these facts in a plausible hypothesis?

REFERENCES

Azuara, M. P. & P. J. Aparicio (1983). In vivo blue-light activation of *Chlamydomonas reinhardii* nitrate reductase. *Plant Physiol.* 71, 286–290.

Abdel-Rahman, M. H. (1982). Photopériodisme chez *Acrochaetium asparagopsis* (Rhodophycées). I. Reponse à une photoperiode de jours courts au cours de la formation de tétrasporocystes. *Physiol. Veg.* 20, 155–64.

Abdel-Rahman, M. H. (1982). II. Influence de l'interruption de la nyctipériode, par un eclairement blanc ou monochromatique, sur la formation des tétrasporocystes. *C. R. Acad. Sc. Paris.* 294, 389–392.

Clayton, M. (1978). Morphological variation and life history in cylindrical forms of *Scytosiphon lomentaria* (Scytosiphonales: Phaeophyta) from Southern Australia. *Mar. Biol.* 47, 394–357.

Dring, M. J. (1967a). Effects of daylength on growth and reproduction of the Conchocelis-phase of *Porphyra tenera*. *J. mar. biol. Ass. U.K.* 47, 501–510.

Dring, M. J. (1967b). Phytochrome in red alga, *Porphyra tenera*. *Nature* 215, 1411–1412.

Dring, M. J. (1984). Photoperiodism and phycology. *Progr. Phycol. Res.* 3, 159–192.

Dring, M. J. & K. Lüning (1975a). A photoperiodic response mediated by blue light in the brown alga *Scytosiphon lomentaria*. *Planta* 125, 25–32.

Dring, M. J. & K. Lüning (1975b). Induction of two-dimensional growth and hair formation by blue light in the brown alga *Scytosiphon lomentaria*. *Z. Pflanzenphysiol.* 75, 107–117.

Dring, M. J. & K. Lüning (1983). Photomorphogenesis of marine macroalgae. In, W. Shropshire & H. Mohr (eds) *Encyclopedia of Plant Physiology, New Series, vol. 16 B*. Springer-Verlag, Berlin, pp. 545–568.

Dring M. J. & J. A. West (1983). Photoperiodic control of tetrasporangium formation in the red alga *Rhodochorton purpureum*. *Planta* 159, 143–50.

Guiry, M. D. & E. M. Cunningham (1984). Photoperiodic and temperature responses in the reproduction of north-eastern Atlantic *Gigartina acicularis* (Rhodophyta, Gigartinales). *Phycologia* 23, 357–67.

Guiry, M. D. & J. A. West (1983). Life history and hybridization studies on *Gigartina stellata* and *Petrocelis cruenta* (Rhodophyta) in the North Atlantic. *J. Phycol.* 19, 474–94.

Kristiansen, A. (1984). Experimental field studies on the ecology of *Scytosiphon lomentaria* (Fucophyceae, Scytosiphonales) in Denmark. *Nord. J. Bot.* 4, 7, 19–24.

Lüning, K. (1980a). Critical levels of light and temperature regulating the gametogenesis of three *Laminaria* species (Phaeophyceae). *J. Phycol.* 16, 1–15.

Lüning, K. (1980b). Control of algal life-history by daylength and temperature. In, J. H. Price, D. E. G. Irvine & W. F. Farnham (eds). *The Shore Environment, Vol. 2, Ecosystems*. Academic Press, London, pp. 915–945.

Lüning, K.(1981a). Light. In, C. S. Lobban & M. J. Wynne (eds). *The Biology of Seaweeds*. Blackwell, Oxford, pp. 326–355.

Lüning, K. (1981b). Photomorphogenesis of reproduction in marine macroalgae, *Ber. Deutsch. Bot. Ges.* 94, 401–417.

Lüning, K. (1981c). Egg release in gametophytes of *Laminaria saccharina*: induction by darkness and inhibition by blue light and U.V. *Br. phycol. J.* 16, 379–393.

Lüning, K. (1986). New frond formation in *Laminaria hyperborea*, a photoperiodic response. *Br. phycol. J.* 21, 269–273.

Lüning, K. & M. J. Dring (1975). Reproduction, growth and photosynthesis of gametophytes of *Laminaria saccharina* grown in blue and red light. *Mar. Biol.* 29, 195–200.

Lüning, K. & M. Neushul (1978). Light and temperature demands for growth and reproduction of laminarian gametophytes in Southern and Central California. *Mar. Biol.* 45, 297–309.

Mayhoub, H. (1976). Cycle de développement du *Calosiphonia vermicularis* (J.Ag.) Sch. (Rhodophycées, Gigartinales). Mise en evidence d'une réponse photoperiodique. *Bull. Soc. Phycol. Fr.* 21, 48.

Maggs, C. A. & M. D. Guiry (1982). Morphology, phenology and photoperiodism in *Halymenia latifolia* Kütz. (Rhodophyta) from Ireland. *Botanica Mar.* 15, 589–599.

Maier, I. (1984). Culture studies of *Chorda tomentosa* (Phaeophyta, Laminariales). *Br. phycol. J.* 19, 95–106.

Müller, D. G. & N. Lüthe (1981). Hormonal interaction in sexual reproduction of *Desmarestia aculeata* (Phaeophyceae). *Br. phycol. J.* 16, 351–356.

Müller, D. G., M. N. Clayton & I. Germann (1985). Sexual reproduction and life history of *Perithalia caudata* (Sporochnales, Phaeophyta). *Phycologia* 24, 467–473.

Oza, R. M. (1976). Culture studies on induction of tetraspores and their subsequent development in the red alga *Falkenbergia rufolanosa* Schmitz. *Botanica Mar.* 20, 29–32.

Powell, J. H. (1964). The life history of a red alga, *Constantinea*. Ph. D. Thesis, Univ. Microfilms Inc., Ann Arbor, Mich., 154 pp.

Rentschler, H. G. (1967). Photoperiodische Induktion der Monosporenbildung bei *Porphyra tenera* Kjellm. (Rhodophyta-Bangiophyceae). *Planta* 76, 65–74.

Richardson, N. (1970). Studies on the photobiology of *Bangia fuscopurpurea*. *J. Phycol.* 6, 215–219.

Rietema, H. & A. W. O. Klein (1981). Environmental control of the life cycle of *Dumontia contorta* (Rhodophyta) kept in culture. *Mar. Ecol. Progr. Ser.* 4, 23–29.

Rueness, J. & P. A. Asen (1982). Field and culture observations on the life history of *Bonnemaisonia asparagoides* (Woodw.) C. Ag. (Rhodophyta) from Norway. *Botanica Mar.* 25, 577–587.

Senger, H. (ed.). (1984). *Blue Light Effects in Biological Systems*. Springer-Verlag, Berlin, 555 pp.

Starr, R. C. (1978). The culture collection of algae at the University of Texas at Austin. *J. Phycol.* 14 (Suppl.), 47–100.

Terry, L. A. & B. L. Moss (1980). The effect of photoperiod on receptacle initiation in *Ascophyllum nodosum*. *Br. phycol. J.* 15, 291–301.

tom Dieck, I. (1987). Non-photoperiodic geographic isolates of the brown alga *Scytosiphon lomentaria*: developmental regulation by temperature and temperature tolerance (in preparation).

Tseng, C. K., K. Z. Ren & C. Y. Wu (1959). (On the discharge of eggs and spermatozoids of *Laminaria japonica* and the morphology of the spermatozoids). *Kexue Tongbao* 4, 129–130 (in Chinese).

NOTES FOR INSTRUCTORS
Irradiance

1. White light field: any cool fluorescent light source will do.
2. For blue or red light, projectors combined with appropriate interference filters may be used, or the boxes can be made with colored plexiglas (or at least plexiglas lids) and irradiated with colored light to give following broad-band spectral light fields, as follows:
   a) Blue light: Philips blue fluorescent lamps TL 40 W/18, combined with 3 mm blue Röhm and Haass Plexiglas; emitted waveband of

lamp-filter combination: 400–520 nm, with peak emission at 445 nm.

b) Red light: Philips red fluorescent lamps TL 40 W/15, combined with 3 mm red Röhm and Haass Plexiglas, No. 501; emitted waveband of lamp-filter combination: 610–700 nm, with peak emission at 660 nm.

If you use other filter material (other types of red plexiglas or colored acetate foils), be sure that no wavelengths shorter than 520 nm are transmitted. Measure filtering material in spectrophotometer or look up manufacturer's data sheets. Care must be taken to exclude all white light during the experiments.

Medium (PES) (Starr 1978)

Enrichment solution:

1. glass-distilled water	100 mL
2. $NaNO_3$	350 mg
3. Na_2 glycerophosphate	50 mg
4. Fe (as EDTA - 1:1 molar*)	2.5 mg
5. PII metals**	25 mL
6. vitamin B_{12}	10 μg
7. thiamine	0.5 mg
8. biotin	5 μg
9. Tris buffer	500 mg

* Dissolve 351 mg $Fe(NH_4)_2(SO_4)_2 \cdot 6H_2O$ and 300 mg Na_2EDTA in 500 mL H_2O. 1.0 mL of this solution = 0.1 mg Fe.

** PII metal mix: to 100 mL distilled water add

H_3BO_3	114 mg
$FeCl_3 \cdot 6H_2O$	4.9 mg
$MnSO_4 \cdot 4H_2O$	16.4 mg
$ZnSO_4 \cdot 7H_2O$	2.2 mg
$CoSO_4 \cdot 7H_2O$	0.48 mg
Na_2EDTA	100 mg

Adjust pH to 7.8. Add 2 mL of enrichment solution per 100 mL filtered seawater.

Additional chemicals and glassware

1. GeO$_2$
2. closed glass vessels, e.g., crystallizing dishes or deep petri dishes
3. culture dishes
4. coverslips
5. microscope with green or red interference filter

Algae

See introduction and Table 28.1

Scheduling

These exercises can be set up in a 3-h laboratory period, but must be continued for several weeks with short periods of observation.

29

Demonstration of sexual pheromones in Laminariales and Desmarestiales

Dieter G. Müller

Fakultät für Biologie der Universität, D-7750 Konstanz,
Federal Republic of Germany

FOR THE STUDENT

Introduction

Interaction between gametes of opposite sex is a well known phenomenon in lower plants. Several steps precede contact and fusion of gametes. Chemical factors induce sexual organs in *Achlya, Volvox* and fern gametophytes (van den Ende 1976). Sexual chemotaxis is especially well documented in marine brown algae, and the chemical identity of attractants has been established in several species (Müller 1981; Maier and Müller 1986). A new phenomenon was discovered recently in the orders Laminariales and Desmarestiales: the existence of spermatozoid-releasing pheromones. Release and attraction of spermatozoids in these two orders is mediated by chemical factors (Lüning and Müller 1978; Müller et al. 1982). Brown algal pheromones are low-molecular olefinic hydrocarbons (Fig. 29.1). They are poorly water-soluble, highly volatile, and persist for days or weeks in seawater, and for years when stored adsorbed or in solution at $-15°C$. Their volatile character can be used for the isolation and bioassays of these compounds. They are present in the air space over a culture with freshly released eggs of any species belonging to the orders Laminariales or Desmarestiales. A suitable adsorbent can accumulate the pheromone and release it for a bioassay with the respective male gametes or gametophytes.

243

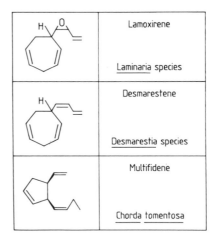

	Lamoxirene
	Laminaria species
	Desmarestene
	Desmarestia species
	Multifidene
	Chorda tomentosa

Fig. 29.1. Structures of the spermatozoid-releasing and attracting pheromones discussed in this experiment.

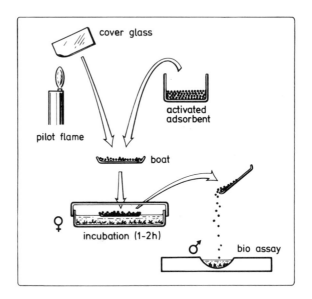

Fig. 29.2. Experimental steps for bioassay.

Procedure

Preparation and exposure of adsorbent (Fig. 29.2). Prepare a boat to hold the adsorbent in open contact with the air space above the female culture as follows: Cut a standard cover glass in half. Heat its corners and edges carefully in a small flame until they curl up slightly. Activate the adsorbent by heating to +150°C for 15 min in an oven. Load the boat with about 20 particles of activated adsorbent and float it for 1–2 h on the surface of the female culture with the cover of the petri dish closed.

Bioassay (Fig. 29.2). Pipette a drop of culture medium with fertile male gametophytes into a depression well. A few adsorbent particles are dropped into this suspension, and the effect observed under a dissection microscope or low-power compound microscope. Beginning about 10 sec after introduction of the pheromone, spermatozoids will be expelled from the antheridia, swim freely and accumulate around the adsorbent particles (Fig. 29.3). A parallel experiment with untreated adsorbent particles will show that they have no effect. Care has to be taken to keep the temperature during the bioassay at 10°–12°C, since warming to room temperature will cause spontaneous release of spermatozoids in many species.

Questions

1. What is the chemical structure of the pheromones demonstrated in the experiment?
2. Why is the pheromone present in the air, while its biological function is obviously in the aqueous milieu?

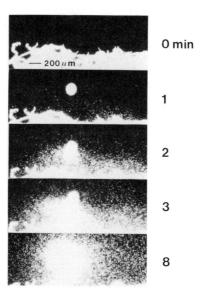

Fig. 29.3. Effect of one adsorbent particle containing lamoxirene on male gametophytes of *Alaria esculenta*. Time-series of dark field photographs.

3. What is the ecological significance of the hormonal systems studied here?
4. How does the pheromone effect the discharge of antheridia?
5. Is release and attraction of spermatozoids effected by the same pheromone?

REFERENCES

Ende, H. van den (1976). *Sexual Interaction in Plants.* Academic Press, London.

Lüning, K. & D. G. Müller (1978). Chemical interaction in sexual reproduction of several Laminariales (Phaeophyceae): Release and attraction of spermatozoids. *Zeitschr. Pflanzenphysiologie* 89 pp. 333–341.

Maier, I., D. G. Müller, G. Gassman, W. Boland, F.-J. Marner & L. Jaenicke (1984). Pheromone-triggered gamete release in *Chorda tomentosa. Naturwissenschaften* 71, 48.

Maier, I., & D. G. Müller (1986). Sexual pheromones in algae. *Biol. Bull.* 170 pp. 145–175.

Müller, D. G. (1981). Sexuality and sex attraction. In, C. S. Lobban & M. J. Wynne (eds) *The Biology of Seaweeds,* Botanical Monographs 17. Blackwell, Oxford, pp. 661–674.

Müller, D. G., A. Peters, G. Gassmann, W. Boland, F.-J. Marner, & L. Jaenicke (1982). Identification of a sexual hormone and related substances in the marine brown alga *Desmarestia. Naturwissenschaften* 69, 290.

Müller, D. G., I. Maier & G. Gassmann (1985). Survey of sexual pheromone specificity in Laminariales (Phaeophyta). *Phycologia* 24 pp. 475–477.

Peters, A. F. & D. G. Müller (1986). Life-history studies – a new approach to the taxonomy of ligulate species of *Desmarestia* (Phaeophyceae) from the Pacific coast of Canada. *Can. J. Bot.* 64 pp. 2192–2196.

Ramirez, M. E., D. G. Müller & A. F. Peters (1986). Life history and taxonomy of two populations of ligulate *Desmarestia* (Phaeophyceae) from Chile. *Can. J. Bot.* 64 pp. 2948–2954.

Starr, R. C. (1978). The culture collection of algae at the University of Texas at Austin. *J. Phycol.* 14: Supplement, pp. 47–100.

NOTES FOR INSTRUCTORS
Adsorbents

Suppliers of gas chromatography accessories offer various materials that are based on porous glass or silica particles with large internal surface area. Such materials have been found to adsorb brown algal pheromones from the air.

In my laboratory, the following products have been used successfully for this purpose:

1. Controlled Pore Glass: irregular particles, nominal pore size 24 nm, mesh size 80–120, 118 $m^2 g^{-1}$
   Suppliers: Sigma GmbH, D 8028 Taufkirchen, FRG; Sigma, PO Box 14508, St. Louis, MO 63178, USA, stock number PG 240–120.
2. Porasil C, globular particles, pore size 20–40 nm, mesh size 100–150, 50 $m^2 g^{-1}$
   Suppliers: Serva, Postf. 105260, D 6900 Heidelberg, FRG, Catalog Nr. 46374.
   Applied Science Laboratories, State College, PA 16801, USA; Applied Science Laboratories Europe, B.V., PO Box 1149, 3260 Oud-Beijerland, Netherlands, catalog Nr. 05893.
3. Spherosil, globular particles, diameter 100–150 μm, pore size 20–40 nm.
   This material, referred to in earlier studies, is no longer on the market and must be replaced by comparable products like those listed above.

Spot plates

"Microtiter" plates with 8 × 12 depressions are offered by laboratory suppliers as a product of Cooke Engineering Company. Similar items may also be used.

Species recommended

The following recommendations are based on our recent work on sexual pheromones in Laminariales and Desmarestiales and are given as a guideline.

Laminariales: An extensive survey showed that all members of the families Laminariaceae, Alariaceae and Lessoniaceae use lamoxirene as a sexual pheromone (Müller et al. 1985). Thus, cross reactions are obtained within all taxa of these three families. The following genera have been found to be suitable for mass cultures of gametophytes and easy formation of antheridia and oogonia under culture conditions: *Laminaria, Alaria, Undaria, Lessonia, Macrocystis*. Some other genera are less easy to propagate vegetatively under laboratory conditions. *Chorda tomentosa*, although monoecious, has been found to produce high amounts of multifidene (Fig. 29.1) as its sexual pheromone (Maier et al. 1984).

Desmarestiales: Best results were obtained with *Desmarestia aculeata*. *D. viridis* is not suitable because its gametophytes are monoecious. Ligulate species of *Desmarestia* from the Pacific are suitable if they are dioecious (Peters and Müller 1986; Ramirez et al. 1986).

Demonstration film

The topic of this experiment is available as an educational film (16 mm, 9 min, English language) with the title "Pheromone Effects in Fertilization of Brown Algae," distributed as Film Nr. C 1424 by: Institut für den Wissenschaftlichen Film, Nonnenstieg 72, D 3400 Göttingen, FRG.

Gametophyte cultures

The experiment requires mature male and female gametophytes of *Laminaria* or *Desmarestia*. It is technically simple and illustrates the presence and activity of sexual pheromones in lower plants. The interval of about 10 sec from application of the pheromone to the release of spermatozoids is the fastest effect of plant hormones known at present. Since it requires long-term preparations, the experiment is best suited for repetition at regular intervals, once the cultures and their handling are established in the laboratory. The initiation of clonal cultures (description follows) can be used as a class project and the cultures can also be used for experiments on induction of sexual reproduction (Exp. 28).

Gametophytes are grown in enriched natural seawater (Provasoli medium, Starr 1978; details follow). Cultures are illuminated for 14 h day^{-1} with a red fluorescent lamp (Philips TL 20 W/15 or similar type at 3 μE m^{-2} sec^{-1}). White light must be excluded. Transfer to fresh medium is done about every two months.

Gametogenesis is induced by transfer to white fluorescent light with increased photon flux density of 10–20 μE m^{-2} s^{-1} at 14 h day^{-1}. Eight to 10 days later, oogonia and antheridia are present, and such cultures can be used for the experiment.

A simple way to maintain gametophyte cultures is to use a normal household refrigerator without deep-freeze section, with one red fluorescent lamp inside. If the ballast is removed from the lamp housing and installed separately outside the refrigerator, a sufficiently stable temperature level of 10°–12°C can be maintained inside.

Since cold water algae are presently not offered by culture collections, steps for isolation of gametophytes from field material are given below. Starting from meiospores of sporophytes collected in the field, the isolation and establishment of unisexual clonal cultures of sufficient size requires roughly half a year.

Isolation of gametophytes from field sporophytes

1. Select fertile thalli from field specimens. In Desmarestiales unilocular sporangia are dispersed on the thalli rather than in prominent sori typical for Laminariales. Make sure that thalli are fertile by examining microscopic sections.
2. With a razor blade, cut a series of small pieces of fertile tissue about 2×2 mm, and subject them to different treatments in order to obtain spore release. With mature fertile tissue, at least one of the three following treatments will usually result in rich spore release:
   a) add one drop of culture medium immediately.
   b) wipe some pieces dry with paper tissue and put them into a plastic petri dish, seal with Parafilm and store overnight in a refrigerator at 4°–10°C.

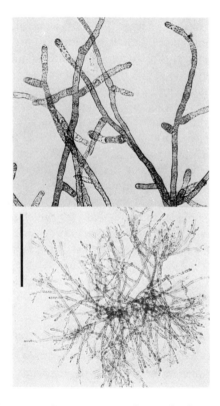

Fig. 29.4. Heteromorphism in gametophytes of a dioecious species of *Desmarestia* grown in red light. Female (top) and male gametophyte (bottom), scale bar 200 μm

c) rinse some pieces in tap water for a few seconds, wipe dry, and store in the refrigerator as under (b). The next day, add one drop of cold culture medium to the thallus fragments from (b) and (c). Confirm spore release with microscopic observation.

3. When meiospores are obtained, pipette small aliquots (one drop of a rich spore suspension) into plastic petri dishes with culture medium and incubate at 10°C under red light.

4. After a few weeks, meiospores develop to filamentous gametophytes. Most members of the orders Laminariales and Desmarestiales are dioecious, with clearly heteromorphic gametophytes: female gametophytes have broader filaments and are darkly pigmented, while filaments of male gametophytes are narrower in diameter, more richly branched and less pigmented (Fig. 29.4). Under the stereo-microscope, select clean male and female gametophytes. Remove them from the substrate with a needle and inoculate one germling into each well of a microtiter spot plate using Pasteur pipettes. This step serves also to eliminate contaminants: diatoms and other algae or protozoa are removed indirectly without the toxic agents (germanium dioxide or antibiotics) frequently used by many phycologists. After inoculation, the spot plate is covered, sealed with Parafilm, and incubated under red light.

   After four to six weeks, the inoculants in the spot plates are examined under the stereo microscope. The large number of trials gives very good chances of obtaining at least a few cultures that are free of contaminants. Select healthy and clean gametophytes. Remove them from the spot plate with a Pasteur pipette and transfer into plastic petri dishes with 5 mL culture medium.

5. At this stage, first attempts at propagation of individual plants by fragmentation can be made: the gametophytes, which grow as small semi-globular tufts, can be fragmented with sterile dissection needles or by using the edge of a Pasteur pipette for cutting the plants into fragments. Further incubation follows in order to increase the biomass of individual gametophyte clones. In later stages, fragmentation of gametophytes can be done by using mild homogenization with a glass tube and a loosely fitting piston.

Using these techniques, clonal male and female cultures will reach the volume to populate a few medium-size petri dishes within about half a year. This will be sufficient material for the experiment.

Generation of "living pictures" by photoreactive microalgae

Bruno P. Kremer

Universität zu Köln, Institut für Naturwissenschaften und ihre Didaktik, Abteilung für Biologie, 5000 Köln 41, Federal Republic of Germany

FOR THE STUDENT

Introduction

Diverse micro-organisms are capable of active movements, which are of fundamental significance for settlement in a suitable habitat or environment. The movements are a response to an array of abiotic factors including chemical, thermal, mechanical and optical signals. Light is one of the major factors and acts as a controlling system particularly (but not exclusively) in photosynthetic organisms.

Three different types of active orientation in the light are currently recognized and distinguished. Phototaxis encompasses all orientated movements that are controlled in relation to the direction of the incident light. Many flagellates show a positive phototaxis toward the light and are thus very efficiently attracted to a light source. In some cases, the response is away from the light, particularly in the case of very high irradiances (negative phototaxis). Photokinesis, on the other hand, is independent of the light direction. In this type of photic response, the motility of the organisms is enhanced or inhibited by changes in light intensity. A positive photokinesis eventually results in a concentration of organisms in darker zones, since statistically they remain for a longer time in the dark and leave the light zones with accelerated speed. Finally, in photophobic reaction, a further quality of response is observed. A rapid temporal or spatial change in light intensity elicits a transient change in motility. Usually the cell stops and then reverses direction.

A peculiar kind of movement, gliding in and out of a mucilaginous sheath, is found with filamentous cyanobacteria. The filaments (trichomes) of *Phormidium* are unable to swim freely through the medium, but instead slowly glide if in contact with their substrate. The mechanics of such a movement, particularly the generation of the thrust forces, are still unknown. However, the general increase of gliding speed with an increase of incident light apparently indicates that an enhancement in photophosphorylating activity is involved on the energetic side of the phenomenon. The fascinating field of photic motility responses in micro-organisms was reviewed by Häder (1979), Nultsch (1980), and Castenholtz (1982).

The photic motility response of *Phormidium uncinatum* has been very well investigated. When a gliding trichome of this cyanobacterium reaches a zone of lower light intensity, it soon stops for a few seconds and then glides back along its original path (photophobic reversal of movement). When light spots are projected in a culture of *Phormidium*, a certain proportion of the trichomes will glide over the border between the darker environment and the light field. Although the sudden increase in light intensity does not trigger a photophobic reaction, the decrease in light intensity during leaving a spot immediately elicits a reversal. Hence, over a couple of hours, more and more of the gliding trichomes become concentrated in the light areas. The light spots thus act as an efficient light trap. This phenomenon forms the background of the following experiments.

Microscopic observation of gliding trichomes

Procedure. Place a small sample of *Phormidium* with some tap water (or culture medium) on a microscope slide and cover with a cover slip. As soon as the trichomes have fixed their mucilage sheaths to the glass surfaces, the gliding movement starts. *Phormidium uncinatum* develops a gliding speed of 150–170 μm min^{-1}. At higher magnification it may be seen that the trichomes also rotate counter-clockwise (relative to the direction of movement).

The formation of tube-like mucilage sheaths may be easily observed if some drawing ink is added as a marker. Dip a dissecting needle into black drawing ink and mix the adhering amount with a drop of water. The resulting blackening is quite sufficient for your further observations. Higher concentrations of ink are toxic, since it often contains phenol. The ink particles mark the mucilage tube very well and show that the sheath collapses when the trichomes have left them. Furthermore, the rotation of the trichomes causes a noticeable torsion of the sheath.

Reversal of the gliding movement can be induced by a sudden decrease of the observation light in the microscope, e.g., by rapidly turning down the lamp power, or by inserting a neutral filter (with a transmission in the range of 10%) in the light path. Within a few seconds (up to 2 min) most of the trichomes stop their gliding and then change the direction of movement.

Spatial differences in available light intensity are best offered by stopping down the condenser of the microscope. The inner edge of the condenser diaphragm should be in focus. After some minutes, the majority of trichomes will have concentrated in the inner light field.

Questions. Measure the velocity of your *Phormidium* trichomes with a calibrated ocular micrometer. Compare the average speed with the average length of the trichomes.

How long does it take until the gliding movement of the filaments (trichomes) stops upon decreasing incident light? How much time is required for the starting with the reversal of movement?

How do trichomes of *Phormidium* respond when entering a light field? Do they speed up? Do they change the direction of movement upon entering or upon leaving the light field?

The trichomes of *Phormidium* comprise a large number of individual cells. What proportion of the trichome (cells) must be in the light (dark) to trigger a photophobic response?

If a photomicroscope equipped with an exposure meter is available, try to determine whether or not the photophobic response is affected by the amount of light decrease/increase. Is there any critical threshold?

Creating living pictures

Procedure. The remarkable tendency of the trichomes to concentrate in light fields can be used to generate patterns in a Petri dish. Harvest a culture of *Phormidium* (about 0.5–1 g fresh weight). Cut the trichomes into short fragments with a razor blade on a glass plate. Boil 0.1 g agar in 100 mL H_2O and cool it to about 40°C. Meanwhile rub a Petri dish with a little glycerol. Mix approximately 20 mL of the 40°C agar with the *Phormidium* suspension and pour carefully into the rubbed Petri dish (avoid air bubbles!). Place the algae in the dark overnight and then illuminate

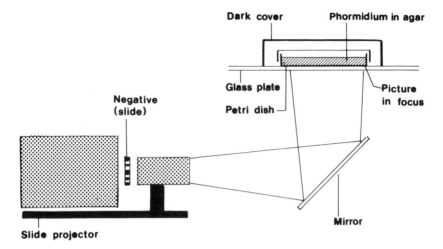

Fig. 30.1. Arrangement for the generation of a 'living picture' within a culture of *Phormidium uncinatum:* The negative (mounted as slide) is projected onto the bottom of a Petri dish. The entire set is placed in a dark room. Distance between slide projector, mirror and culture dish not to scale.

Fig. 30.2. Example of a *Phormidium* photograph showing a roller coaster. The algograph or bacteriograph developed in 12 h and was fixed by drying out. (Courtesy Dr. D. P. Häder)

them through the pattern of a carboard cut-out (Fig. 30.1). After several hours (check carefully every 2 h) the trichomes have concentrated in the respective light traps given by the patterns within the cardboard.

Instead of triangles, stars, holes, etc., cut into a black cardboard, a photographic negative, mounted as a slide, can be focused exactly onto the bottom of the Petri dish (Fig. 30.1). Since the trichomes differentially respond to gradual changes in transmission, halftone pictures can be generated in the Petri dishes: the photophobic reaction of *Phormidium* results in the positive picture (Figs. 30.2 and 30.3). The development of such a living picture takes about 10–12 h continuous illumination. The whole range of possible motifs including landscapes, architecture or portraits could be tried.

It is possible to preserve such positive pictures. After about 12 h exposure, the cover of the Petri dish is carefully removed and the agar allowed to dry out, while the bottom of the dish is still illuminated with the focused picture. Thereby the trichomes are simply fixed in position in the drying agar. The glycerol with which the dish was rubbed prevents splitting of the agar during this procedure. The dried *Phormidium* photographs can be maintained for many years. However, they tend to bleach out soon when fully exposed to sunlight.

Fig. 30.3. Another example of bacteriography using photoreactive *Phormidium*. The still unfixed picture in the Petri dish shows a street with framework houses in a small German town. (Courtesy Dr. D. P. Häder)

Questions

1. What is the minimum exposure time for the generation of a living picture using *Phormidium?*
2. Why should the trichomes used in this experiments be rather short? What is the corresponding technical feature of a photographic film?
3. Does the contrast of the projected negative affect the exposure time for the development (generation) of the living positive?
4. Project a multicolored slide onto the bottom of the Petri dish. What is the response of the trichomes? Is it possible to generate the equivalent negative of the color slide? What do the results mean with respect to the spectral sensitivity of *Phormidium?*

Motility of Euglena

Procedure. A similarly impressive experiment on photic motility response, based on positive phototaxis, can be performed using a suspension of *Euglena gracilis.* This organism is easily grown in a mineral medium enriched with a few organics (see following).

A light green, not too dense suspension of *Euglena* cells is poured into a glass cuvette with thin, parallel sides (volume about 100–250 mL) (Fig. 30.4). The cuvette is darkened by a black cover. One side of the cover carries simple geometric patterns or even translucent letters such as the

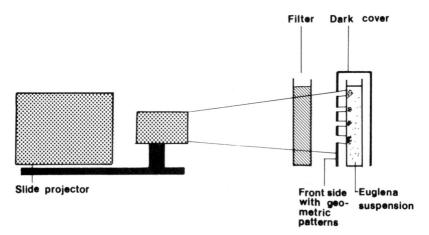

Fig. 30.4. Arrangement for the generation of geometric patterns or lettering in a *Euglena* suspension. The front side of the cover must be placed very close to the cuvette in order to prevent too much light from scattering. A second cuvette placed in front of the suspension (optionally filled with colored liquid) must be used to prevent warming up of the cell suspension under the dark cover. Distance between slide projector and cuvettes not to scale.

series EUGLENA. This arrangement is illuminated by a slide projector for 1–3 h. Due to the positive phototactic response of *Euglena,* the cells again concentrate behind the translucent regions of the cover. When the cover is removed, a positive (= dark green) picture has been generated. Eventually, the *Euglena* suspension is able to 'write' its generic name.

The same arrangement allows further experiments on the spectral sensitivity of the photic motility response. Use chromatic light (blue, green, red) instead of white light by the application of appropriate filters of a defined transmission range (interference filters 5 × 5 cm used as slides in the projector). Alternatively, you can prepare your own filters: Glass cuvettes (about 10 × 10 × 3 cm) are used, filled with (1) a dense, green *Euglena* suspension (providing green light), (2) 1 M $CuSO_4$ (for blue light), and (3) 1 M $K_2Cr_2O_7$ (for red light).

Questions

1. How long does a *Euglena* suspension require for generating a pattern by phototaxis?
2. How stable is the generated picture after lifting the dark cover?
3. What did you find about the spectral sensitivity of the photic motility response by *Euglena*? Are there any recognizable relationships to the pigmentation of the cells? How do the results from this experiment compare with the findings from *Phormidium?*

REFERENCES

Castenholtz, R. W. (1982). Motility and taxes. In, N. G. Carr & B. A. Whitton (eds) *The Biology of Cyanobacteria,* Blackwell Scientific Publications, Oxford, pp. 413–439.

Häder, D. P. (1979) Photomovement. In, W. Haupt & M. E. Feinleib (eds) *Encyclopedia of Plant Physiology, Vol. 7, Physiology of Movements,* Springer, Berlin, pp. 268–309.

Nultsch, W. (1980). Photomotile responses in gliding organisms and bacteria. In, F. Lenci & G. Colombetti (eds) *Photoperception and Sensory Transduction in Aneural Organisms,* Plenum Press, New York, pp. 69–88.

NOTES FOR INSTRUCTORS

Equipment

(per group of students)

1. microscope (photomicroscope with exposure meter optional)
2. slide projector
3. mirror (ca. 10 × 10 cm)
4. dissecting needles, razor blades

5. slide-size interference filters (5 × 5 cm) for providing monochromatic light of approximately 460, 550 and 650 nm.

Glassware

(per group of students)

1. Petri dishes (glass or disposable), standard size (about 10 cm diameter) (2–3)
2. Pasteur pipettes
3. microscope slides, cover slips
4. glass cuvette, ca. 10 × 10 × 3 cm or similar size with thin, parallel sides (3 cuvettes of this size are required, if no special optical filters are available)

Organisms

Phormidium uncinatum Gomont may be obtained from the following sources:

Sammlung von Algenkulturen, Pflanzenphysiologisches Institut der Universität, Nikolausberger Weg 18, D-3400 Göttingen (= SAG): Culture No. 81.79

Culture Collection of Algae, Department of Botany, The University of Texas at Austin, Austin, Texas 78712, U.S.A. (= UTEX) offers no *Phormidium uncinatum*, but *Phormidium foveolarum* (Culture No. 427) and *Phormidium* spp. (Culture No. 595), which appear equally well suited.

The Microbial Culture Collection, The National Institute for Environmental Studies (NIES), Yatabe-cho, Tsukuba-gun, Ibaraki 305, Japan, also offers *Phormidium foveolarum* (NIES-32) and *P. tenue* (NIES-30)

Euglena gracilis Klebs may be obtained from

NIES Strain No. 47 and 48
SAG Strain No. 1224/5.1
UTEX Strain No. 367

Culture

Phormidium is cultured in the following medium:

0.8 g KNO_3
0.2 g $K_2HPO_4 \cdot 3H_2O$
0.2 g $Ca(NO_3)_2 \cdot 4H_2O$ made up to 1000 mL with distilled water
0.2 g $MgSO_4 \cdot 7H_2O$
3.5 g agar

After boiling this mixture, about 20 mL are put in a sterilized Petri dish and covered with a membrane filter (90 mm diameter, about 0.2 μm pore width). After cooling and setting of the medium, the center of the surface is inoculated under the membrane with *Phormidium* and the culture placed at room temperature under low light condition (about 10 μE m^{-2} s^{-1}, equivalent to approximately 500 lux). Avoid direct sunlight. After five to seven days, the cyanobacteria can be harvested. The harvest is greatly facilitated by the membrane filters, as these are accepted by the trichomes as an auxiliary substratum.

The cultures on membrane filters in Petri dishes are ready for experimentation within one week. At that time they comprise 0.5–1 g fresh weight each.

Euglena is cultured in the following medium:

> 0.2 g KNO$_3$
> 0.02 g (NH$_4$)$_2$HPO$_4$
> 0.01 g MgSO$_4$·7H$_2$O
> 20 mL CaSO$_4$, saturated solution   made up to 1000 mL
> 1.0 g Na-acetate   with distilled water
> 1.0 g beef extract
> 2.0 g bacto-tryptone
> 2.0 g yeast extract

Cultures are best grown in 250 mL or 500 mL Erlenmeyer flasks with continuous shaking (rotating shaker) at 25°–27°C and about 40 μE m^{-2} s^{-1} (equivalent to about 1000 lux) with 12:12 h light:dark. Stock cultures can also be maintained in Erlenmeyer flasks on a window sill without direct sunlight. Cultures growing logarithmically are ready for experimentation after ca. one week.

Scheduling

The students should be provided with ready cultures (started one week before the experiments). The scheduling of the experiments could be as follows:

– microscopic examination	1–2 h
– generation of patterns or pictures	*Phormidium* 10–15 h
	Euglena 1–3 h
– preserving of *Phormidium* pictures	overnight (12 h)
– experiments with chromatic light	2–3 h
– discussion	1 h

31

Nucleo-cytoplasmic interactions in cultured *Acetabularia*

Horst Bannwarth

Universität zu Köln, Institut für Naturwissenschaften und ihre Didaktik, Abteilung für Biologie, 5000 Köln 41, Federal Republic of Germany

FOR THE STUDENT

Introduction

Acetabularia (Fig. 31.1) is a green alga (Chlorophyta, Siphonocladales or Dasycladales, Dasycladaceae) living in the shallow lagoons or coral reefs of tropical and subtropical marine waters. Members of the genus *Acetabularia* represent the largest single-celled organisms that keep their unicellular and uninuclear state during the whole vegetative growth and morphogenesis (Fig. 31.2). A fully grown *Acetabularia* cell typically consists of the apical species-specifically shaped cap with cap-rays arranged like the petals of a daisy, the long slender cylindrical stalk and the root-like anchoring basal process, the branched rhizoid. One of the rhizoidal branches contains the single enormous nucleus (Gibor 1966; Puiseux-Dao 1970).

Acetabularia combines a number of unusual features that make it a desirable and favorite experimental object for cell biologists (Schweiger and Berger 1979): The large size, the polar organization, the distinct life cycle, the species-specific morphogenesis, and not least the astonishing ability for regeneration of nucleate and even anucleate cell fragments.

260

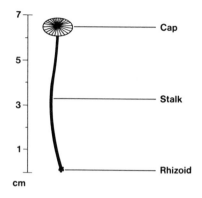

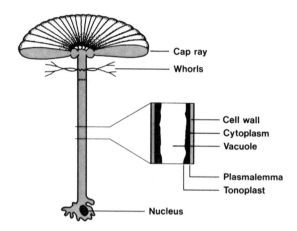

Fig. 31.1. A single *Acetabularia* cell.

The full grown cell (top) consists of the cap, the stalk and the rhizoid containing the nucleus. The schematic representation below shows the characteristic features of a plant cell: A thick cell wall encloses the green cytoplasm and a giant vacuole.

Other advantages result from these features:

Enucleation and fragmentation of cells by surgical means is simple.

Nucleate cells can be kept growing indefinitely by repeated stalk amputation.

Recombination, hybridization and cell fusion including nuclear transfer and exchange can be made by transplantation of cell parts.

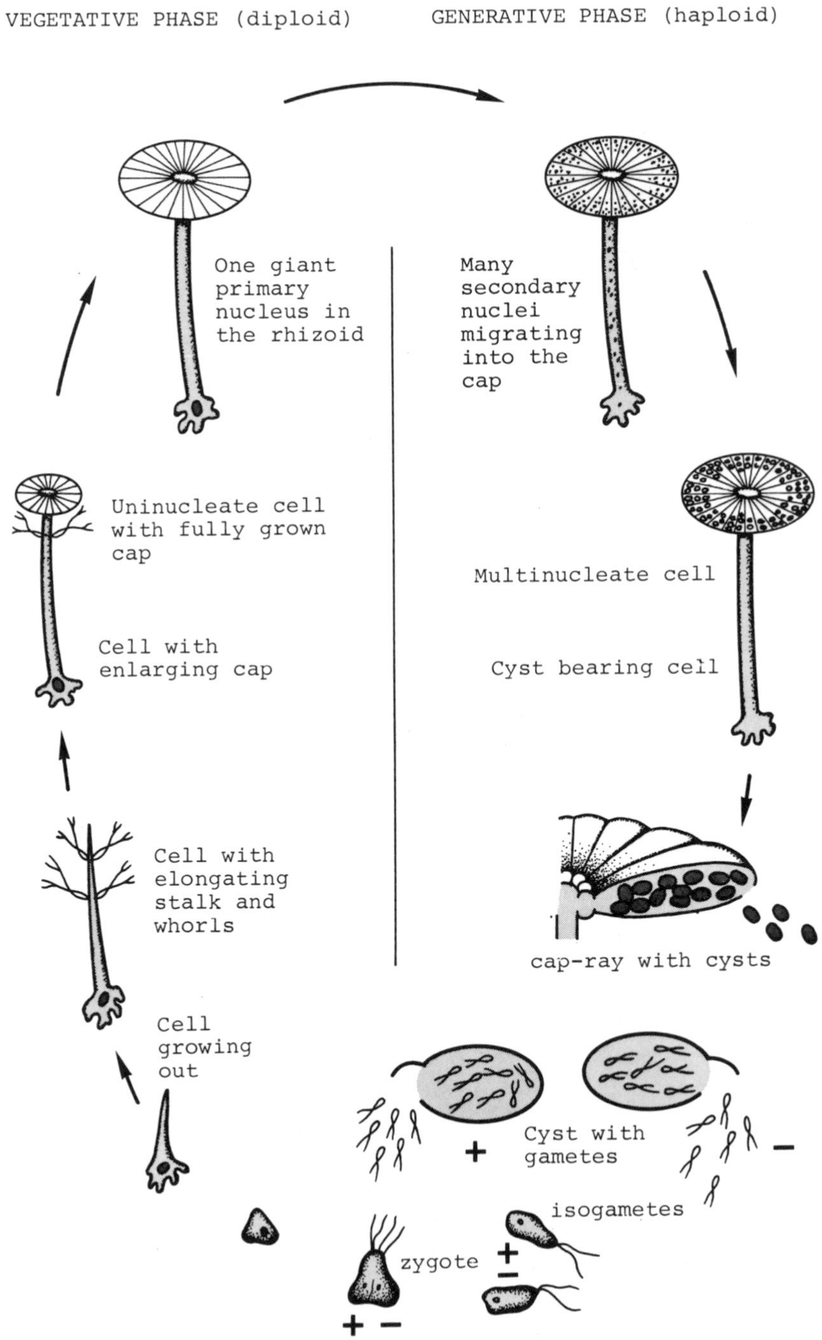

VEGETATIVE PHASE (diploid)   GENERATIVE PHASE (haploid)

One giant
primary
nucleus in
the rhizoid

Many
secondary
nuclei
migrating
into the
cap

Uninucleate cell
with fully grown
cap

Cell with
enlarging cap

Multinucleate cell

Cyst bearing cell

Cell with
elongating
stalk and
whorls

Cell
growing
out

cap-ray with cysts

+ Cyst with
   gametes −

isogametes

zygote +
        −

+ −

Fig. 31.2. The life cycle of *Acetabularia*.

Single nuclei can be isolated by squeezing out the rhizoid protoplasm and can be implanted into anucleate or nucleate cell fragments (Berger and Schweiger 1980).

Growth and morphogenesis can be manipulated by light, chemicals, surgery and ligation (Bannwarth and Schweiger 1983; Bonotto, Cinelli and Billiau 1985).

The pioneering work on *Acetabularia* was done by Joachim Hämmerling in the 1930s mainly on the species *Acetabularia acetabulum = mediterranea* (Hämmerling 1963). His classical experiments provided basic information for an early understanding of nucleo-cytoplasmic relationships.

More recently, a combination of biochemical, biophysical and cell biological methods have revealed that enucleated cells and anucleate apical cell fragments are not only able to grow and to differentiate, showing very impressive morphogenesis, but also to synthesize various metabolites, macromolecules, proteins, enzymes and even RNA and DNA (Bonotto, Lurquin and Mazza 1976). They sustain respiration, photosynthesis, and circadian rhythm for months (Schweiger and Berger 1979).

Morphogenesis, especially the formation of the species-specific cap, is the visible expression of genetic information originally derived from the cell nucleus. Nuclear genes are transcribed early and the genetic instructions are transported into the cytoplasm. Accumulated in the tip of a young *Acetabularia* cell and arranged in an apical-basal gradient, they are stored until they are needed for differentiation. These nuclear instructions, probably mRNA, have been called "Morphogenetische Substanzen (= morphogenetic substances)" by Hämmerling as a consequence of their materiality and significance (Hämmerling 1963).

Nuclear DNA synthesis, the replication of the nuclear genome and the multiplication of nuclei, are associated with the generative phase. It begins after the cap has attained its final size, with the first division of the primary cell nucleus. Several observations recently showed that meiosis takes place at the first nuclear division (Koop 1979). More than ten rapid mitotic cycles of karyokinesis follow. As a result, 10^8, or more, haploid secondary nuclei may arise in less than three weeks. The generative phase ends with the production of physiologically different $(+)$ and $(-)$ biflagellate isogametes. The zygote is created by fusion of a $(+)$ gamete with a $(-)$ gamete (Schweiger and Berger 1979).

The following series of experiments demonstrates the nuclear influence on the cell morphogenesis and conversely the effect of changes of the cell on the nuclear developmental cycle. The methods will include cell surgery

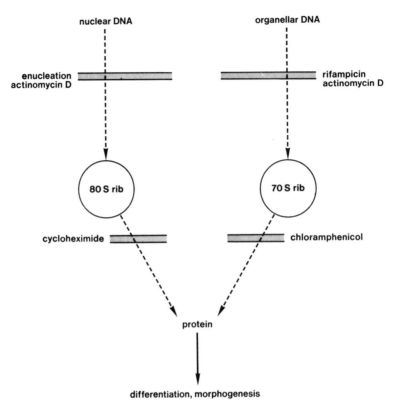

Fig. 31.3. Effects of inhibitors and enucleation on RNA and protein synthesis as well as on cell differentiation.

and the use of inhibitors (Bonotto et al. 1976; Schweiger and Berger 1979). Enucleation and the effects of inhibitors allow us to decide which level – transcription or translation, RNA or protein synthesis – is involved in morphogenesis and differentiation (Fig. 31.3). The inhibitors are as follows:

Inhibitor	Inhibition
actinomycin D	RNA synthesis, transcription of DNA in eukaryotic nuclei and of prokaryotic DNA
cycloheximide	protein synthesis, translation on 80S ribosomes of the cytosol in eukaryotes.
chloramphenicol	protein synthesis, translation on 70S ribosomes in organelles (chloroplasts, mitochondria) and in prokaryotes (bacteria, cyanobacteria).

Further techniques of isolation and implantation of single nuclei, as well as procedures for the demonstration of the specific influence of the nucleus on isoenzyme properties, are described or cited elsewhere (Berger and Schweiger 1980).

Procedures

Cell fragmentation (Fig. 31.4). Select 120 cells of *Acetabularia acetabulum* at the developmental stage just prior to cap formation (stalk length 4–6 cm) from cultures for the experiment. Tie off 100 cells with a thread in the middle and cut them in halves by scissors or a razor blade in a petri dish with culture medium. Ligation prevents a major loss of cytoplasm following surgery. Do not ligate nucleate fragments. Retain 20 cells as untreated controls (AB).

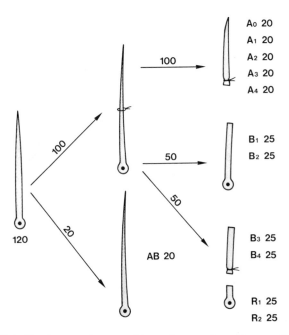

Fig. 31.4. Simple experiments with nucleate and anucleate cell fragments of *Acetabularia*.

Collect and subdivide the *anucleate apical* cell parts into five groups of 20 pieces. Transfer each of them to 100 mL of the following solutions:

        1. Culture medium without additions (A_0)

        2. Ribonuclease (100 mg L^{-1})   (A_1)

   3. Actinomycin D (10 mg L^{-1}) (A_2)

   4. Cycloheximide (1 mg L^{-1}) (A_3)

   5. Chloramphenicol (100 mg L^{-1}) (A_4)

CAUTION, inhibitors are poisonous!

 Mark the bases of the petri dishes, not the covers, with felt pen.

 Tie off 50 of the *nucleate basal* cell parts near the base, and enucleate by amputation of the rhizoid with scissors (Fig. 31.4). Again, only the anucleate fragments are ligated at the basal end.

 Transfer 25 cell fragments of each group to 100 mL of culture medium or actinomycin D solution, as follows:

   1. Nucleate basal halves in culture medium (B_1)

   2. The same, but with actinomycin D (10 mg L^{-1}) (B_2)

   3. Anucleate basal stalks in culture medium (B_3)

   4. The same, but with actinomycin D (10 mg L^{-1}) (B_4)

   5. Nucleate rhizoids in culture medium (R_1)

   6. The same but with actinomycin D (10 mg L^{-1}) (R_2)

Keep all of the dishes of cell fragments under culture conditions.

 Count the number of caps formed on the apical parts in each group of 20 pieces after 3–4 weeks. In which cases is the cap formation suppressed? How would you explain the results in molecular biological terms?

 Measure the average length of 25 basal parts in each treatment using millimeter graph paper under the petri dishes. Repeat the measurement after 3–4 weeks. In which cases are the stalks elongated? In which groups do whorls or caps appear? What can you conclude about the nuclear influence on cell growth and differentiation from this experiment?

Grafting: intraspecific nuclear exchange. (Figs. 31.5, 31.6). Select 40 cells with caps 4–6 mm in diameter. The stalks should have a deep green color so you can be sure that the cells are mononucleate and the translocation of the green cytoplasm from the stalk into the cap did not yet begin (Fig. 31.2). Select also 40 younger cells with a stalk length of 3–4 cm and which did not start to form a cap. Retain 20 old and 20 young cells as controls and exchange the nuclei between 20 old and 20 young cells by grafting. Perform cell surgery in large flat petri dishes, the bottom covered with culture medium. Make all cuts oblique, because this facilitates the following combination of cell parts (Fig. 31.5).

 First, cut off with scissors the rhizoids with a short stalk piece, 5 mm in length.

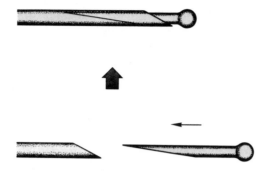

Fig. 31.5. Method of grafting cell parts of *Acetabularia*.

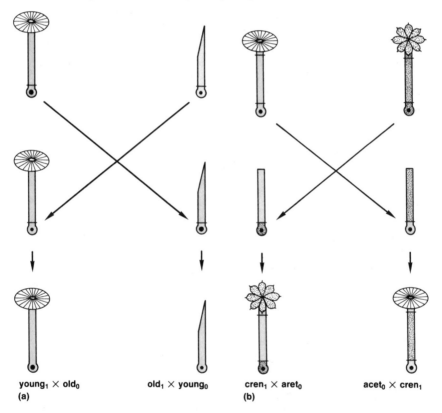

young$_1$ × old$_0$ old$_1$ × young$_0$ cren$_1$ × aret$_0$ acet$_0$ × cren$_1$

(a)   (b)

Fig. 31.6. Intraspecific and interspecific grafts with *Acetabularia*. (a) Intraspecific grafts provide information about the influence of the cytoplasm on the nucleus. The old nucleus attains the features of a young nucleus if it is transferred into a young cell, the young nucleus changes rapidly to an old nucleus inside an old cell. (b) Interspecific grafts show that cell morphogenesis is specified by the nucleus.

Join cell parts using fine normal forceps with one hand and bent forceps with the other hand. Push the end having the smaller diameter into the larger one until the stalk cell walls overlap by a few millimeters.

Combine the nucleate basal rhizoid fragments of young cells with the anucleate cap bearing apical fragments of old cells (young$_1$ $\times$ old$_0$) and the nucleate basal rhizoid fragments of old cells with the anucleate apical fragments of young cells (old$_1$ $\times$ young$_0$) (Fig. 31.6 a).

Transport the grafts without shaking to continuous light for 2 days. The controls are returned to normal culture conditions (light-dark cycle). Grafts showing contractions and discontinuities or the two parts not grown together after one day will not recover and should be discarded.

After 2 days transfer the grafts to the light-dark cycle and culture with the controls for 2 months. Change the medium after one month.

Estimate the number of days until cyst formation and note the average value for the time of appearance of the first five caps with cysts in each group. Cysts can be seen inside the caps with the naked eye.

Is the speed of cell development changed by the age of the nucleus? What can you conclude about the control of the cytoplasm on the nucleus since cyst formation indicates a stage-specific time point of the nuclear cycle (Fig. 31.2)? Is the nuclear development closely related to the life cycle of the cell?

Grafting: interspecific nuclear exchange. Select 50–100 mononucleate cells of *Acetabularia acetabulum* and the same number of *Acetabularia crenulata*. Cells should have small caps, 3–5 mm in diameter. Remove caps with scissors from all the cells selected. Cut off the basal 5 mm of 20 cells of each species, making oblique cuts (Fig. 31.5). Prepare grafts, 20 acet$_1$ $\times$ cren$_0$ and 20 cren$_1$ $\times$ acet$_0$, by combining the nucleate basal fragment of one species with the anucleate middle stalk fragment of the other species under the dissecting binocular microscope (Fig. 31.6 b). Keep 20 cells without caps of each species as controls. Handle grafts and controls as described for intraspecific grafts. Caps should develop on the grafts after 4–6 weeks. Change the medium after 1 month.

Note whether the cap which forms is acet or cren type and compare with the caps regenerated on the nucleate controls. Note in which cases cysts form after 6–8 weeks.

Which morphological markers of the caps formed after the nuclear exchange seem to be determined by the nucleus and which by the cytoplasm? What can you say about the coding site of genes for the species-specific morphogenesis? Propose an experiment to exclude the possible influence of the rhizoid cytoplasm, which is transferred together with the nucleus. How can you explain the fact that cyst formation is often disturbed after the exchange of the cell nucleus in hybrid cells?

REFERENCES

Bannwarth, H. & H.-G. Schweiger (1983). The influence of the nucleus on the regulation of the dCMP deaminase in *Acetabularia. Cell Biol. Int. Rep.* 7, 859–867.

Berger, S. & H.-G. Schweiger (1980). *Acetabularia:* techniques for study of nucleo-cytoplasmic interrelationships. In, E. Gantt (ed.) *Handbook of Phycological Methods.* Cambridge University Press: Cambridge, pp. 47–57.

Bonotto, S., F. Cinelli & R. Billiau (1985). *Acetabularia 1984.* Belgian Nuclear Center, C.E.N.-S.C.K., Mol, Belgium.

Bonotto, S., P. Lurquin & A. Mazza (1976). Recent advances in research on the marine alga *Acetabularia. Adv. Mar. Biol.* 14: 123–250.

Gibor, A. (1966). *Acetabularia:* A useful giant cell. *Sci. Amer.* 215, 118–124.

Hämmerling, J. (1963). Nucleo-cytoplasmic interactions in *Acetabularia* and other cells. *Ann. Rev. Plant Physiol.* 14, 65–92.

Koop, H.-U. (1979). The life cycle of *Acetabularia* (Dasycladales, Chlorophyceae): A compilation of evidence for meiosis in the primary nucleus. *Protoplasma* 100, 353–366.

Puiseux-Dao, S. (1970). *Acetabularia and Cell Biology.* Logos Press, London.

Schweiger, H.-G. & S. Berger (1979). Nucleo-cytoplasmic interrelationships in *Acetabularia* and some other Dasycladaceae. *Intern. Rev. Cytol. Suppl.* 9, 11–44.

NOTES FOR INSTRUCTORS

Materials

1. binocular dissecting microscope
2. fine dissection iridectomy scissors, surgical, bent (Weckerschere, Dosch, Heidelberg)
3. fine normal forceps (Dumoxel Nr. 5, Balzers, Wiesbaden)
4. bent curved forceps (Dumoxel Nr. 7, Balzers, Wiesbaden)
5. Petri dishes, 5.5 cm height and 10 cm in diameter, for culturing
6. Petri dishes, normal
7. Petri dishes, flat, larger in diameter
8. culture medium (see following)
9. inhibitor solutions (see Procedures)
10. felt pen
11. thread, cotton or silk
12. graph paper (millimeter scale)

Materials can be purchased from Carolina Biological Supply Company, 2700 York Road, Burlington, North Carolina 28215; Balzers Hochvakuum, Siemensstr. 11, D-6200 Wiesbaden; Dosch, Bergheimer Str. 9, D-6900 Heidelberg. Inhibitors and ribonuclease, commercial, crystalline, may be obtained from Boehringer, Calbiochem, Serva, Sigma, Worthington Biochemical Co. and others.

Algae

Acetabularia acetabulum
Acetabularia crenulata

The experiments need cultured, uncalcified *Acetabularia* cells. Field collected material will not work.

Acetabularia can be cultured from field collected material (Puiseux-Dao 1970; Berger and Schweiger 1980) but if it is possible, the best way to get *Acetabularia* is to receive them from a research institute. Addresses of institutes working with *Acetabularia* may be obtained from Prof. Dr. S. Bonotto, secretary of the International Research Group on *Acetabularia* (IRGA), Belgian Nuclear Center, C.E.N.-S.C.K., B-2400 Mol, Belgium, Boeretang 200, or from the author.

Culturing

Often, a chemically undefined medium such as the traditional "Erd-Schreiber" medium is used. Recently, good results have been obtained with synthetic media as well as with defined simplified supplemented synthetic seawater (Berger and Schweiger 1980; Bonotto et al. 1985: 54–55) and large scale culture (Bonotto et al. 1985: 319–325). Axenic cultures are not necessary for the experiments presented here.

Acetabularia cultures are grown at a 12:12 hour light-dark cycle, 40–50 μE m^{-2} s^{-1} and temperatures between 18–25°C. The optimum temperature for *Acetabularia acetabulum* is near 20°C, for *Acetabularia crenulata* near 25°C. Fifty–100 single cells can be kept in a large petri dish (5.5 cm in height and 10 cm in diameter) containing 150 mL medium. The medium should be changed monthly.

Scheduling

Studies of growth, morphogenesis and differentiation on *Acetabularia* need time. One should calculate, therefore, an interval of 1–2 months between the experiments and the evaluation. If it is possible to record daily or weekly, the results may also be presented as curves of stalk length or percentage of caps versus time.

Students can work in groups of three to four, with each group tackling a different experiment. The time for student work should be at least 2 h, better 3 h, plus at least 1–2 h for the evaluation and discussion of the experiments.

Comments

Cell biological research on *Acetabularia* has developed very rapidly in the last few years. Molecular genetics and modern cytology have made great progress. Most of the techniques, however, are too complicated and sophisticated for the students or too expensive. On the other hand, since only a small selection of simple experiments can be offered here, students should be encouraged to continue cell research on *Acetabularia*. Isolation, observation and transfer of the giant cell nucleus (Berger and Schweiger 1980) is a further step, but needs some exercise because it is difficult sometimes to discover the colorless nucleus inside the green cytoplasm squeezed out of the rhizoid. Enucleation, grafting and inhibitor use can also be employed for further studies concerning regulatory aspects (Bannwarth and Schweiger 1983). Other experiments can be done, such as the preparation of crude extracts for enzyme assays or the electrophoretic separation of malate dehydrogenase isoenzymes to demonstrate the specific nuclear coding for proteins (Berger and Schweiger 1980).

32

Regeneration of flagella by green flagellates after experimental amputation

G. I. McFadden

Plant Cell Biology Centre, School of Botany, University of
Melbourne, Parkville, 3052 Vic., Australia

I. B. Reize

Westfälische Wilhelms-Universität, Botanisches Institut,
Schlossgarten 3, D-4400 Münster, Federal Republic of
Germany

M. Melkonian

Westfälische Wilhelms-Universität, Botanisches Institut,
Schlossgarten 3, D-4400 Münster, Federal Republic of
Germany

FOR THE STUDENT

Introduction

The eukaryotic flagellum (cilium or undulipodium) is a homologous organelle common to both plants and animals. In flagellates, a primary function of flagella is the generation of motion through the beating action of
the organelle. Protistan flagella are also involved in food gathering, behavioral responses and cell recognition (Amos and Duckett 1982).

A flagellum is a cylindrical extension of the plasma membrane enshea-
thing a collection of cross-linked longitudinal elements (axoneme) embed-

ded in a fluid matrix. The longitudinal elements comprise a central pair of microtubules (hollow fibers composed of the protein tubulin) plus nine outer microtubular doublets (see Fig. 32.1) that attach to the basal body within the cell. The outer doublets are cross-linked by a series of dynein arms (see Figs. 32.1 and 32.2). Each arm contains several polypeptides having ATPase activity (Gibbons and Rowe 1965; Gibbons 1983). Dynein ("force protein") is the mechanochemical transducer, using power from

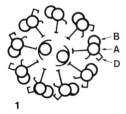

Fig. 32.1. Schematic cross-section of axoneme (viewed from the flagellar base), showing central pair microtubules and nine outer doublets (A and B tubules) with attached dynein arms (D). Radial spokes cross-link the doublets to the central pair. One dynein arm is typically absent in green algal flagella.

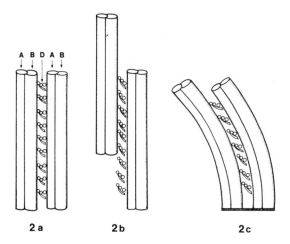

Fig. 32.2. Diagrammatic representation of doublet sliding induced by dynein arms (D) "walking" along the B tubule. a) stationary phase; b) if the doublets were not cross-linked then the left doublet would slide toward the flagellar tip (this actually occurs when the doublet linkages are experimentally removed by specific proteolytic digestion); c) when the doublets are cross-linked and fixed at one end, sliding creates bending.

ATP hydrolysis to "walk" along the microtubules, effectively causing the doublets to slide with respect to one another (Satir 1974; Satir 1984; Satir and Ojakian 1979; Stebbings and Hyams 1979). Since the doublets are attached to the flagellar membrane, the basal body, and cross linked to the central pair microtubules by radial spokes (see Fig. 32.1 and also Ringo 1967), the force generated causes the flagellum to bend (see Fig. 32.2). A useful analogy is that of a telephone book in which sliding of the pages – attached to each other by the binding – causes the book to bend.

Flagella exhibit different beat modes associated with certain behavioral responses, for instance phototaxis (Melkonian and Robenek 1984). Many green flagellates swim forwards with a breast stroke, but can reverse (avoidance reaction) by changing to a symmetrical undulation of the flagella (see Fig. 32.3, and Ringo 1967).

When cells divide, new flagella are formed. In many wall-less green algae, the old flagella are retained and new flagella develop from newly formed basal bodies. In some walled flagellates, like *Chlamydomonas*, the flagella are resorbed prior to division and new flagella regenerate from both old and newly formed basal bodies. In *Tetraselmis*, a scaly green flagellate with a cell wall, the flagella are shed by autotomy before division, and new flagella are produced by the daughter cells. Assembly of flagella from the basal bodies as templates is a most interesting example of organellar morphogenesis, and considering the ubiquitous distribution of flagella – even present in human tissues – information about flagellar biogenesis is important in understanding general cellular developmental processes.

With the discovery that flagella amputated by osmotic, pH, or mechanical trauma were rapidly regenerated from the stumps (Rosenbaum and Child 1967), a system was revealed by which the morphogenesis of the organelle could be experimentally dissected. Most of these studies have

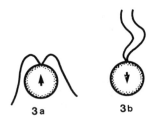

Fig. 32.3. The two common beat modes of green flagellates: a) breast stroke in forward motion (arrow); b) symmetrical flagella undulation (avoidance reaction).

concentrated on the biflagellate *Chlamydomonas reinhardtii* (e.g., Remillard and Witman 1982; Rosenbaum et al. 1969; Witman 1975). Immediately after amputation, the flagellar stubs begin to elongate, growing back to the original length in ca. 1 h (Rosenbaum and Child 1967). If cytoplasmic protein synthesis is inhibited by addition of cycloheximide, the flagella only grow to a portion of the original length and cease regenerating (Rosenbaum et al. 1969). Upon removal of the protein synthesis inhibitor, the flagella resume regeneration (Rosenbaum et al. 1969). These data indicated that a pool of flagellar precursors that can be mobilized for flagellar growth exists in the cell body of *Chlamydomonas* (Rosenbaum et al. 1969). When the regenerating cells were supplied with radio-labeled amino acids, the incorporation of the label into the developing flagella indicated that the axoneme grows mainly by addition of protein at the distal end of the flagellum (Remillard and Witman 1982; Rosenbaum et al. 1969; Witman 1975).

The flagellar membrane and associated extracellular components are thought to be synthesized by the endomembrane system and inserted into the plasma membrane at the base of the developing axoneme by fusion of vesicles with the plasma membrane (Bouck and Chen 1984). Many algae have regularly shaped organic particles (scales) on the flagellar surface (Moestrup 1982), and the production and deployment of these scales have been studied in cells regenerating experimentally amputated flagella (McFadden and Wetherbee 1985; McFadden and Melkonian 1986). These experiments have shown that the Golgi apparatus is also activated by deflagellation, and after a short lag phase begins producing up to 300 scales per minute (McFadden and Melkonian 1986).

Intriguing questions about flagellar development that have not as yet been answered are: (1) what sort of mechanism initiates the assembly of the flagellar precursors, (2) what signal stimulates the precursor synthesis machinery, (3) how do the many hundreds of different macromolecules comprising a flagellum become assembled into the complex biochemical machinery that is able to produce movement, (4) how does the cell detect when the flagella are regenerated to an adequate length, and (5) how are the synthesis signals rescinded?

In this experiment you will amputate the flagella of an algal flagellate and observe their regeneration. The flagella are sheared off mechanically (near the point of emergence) by drag forces created when the cells are rapidly accelerated between the walls of a tissue homogenizer. The rate of flagellar growth, and possible effects of temperature, illumination, and protein synthesis inhibition will be investigated.

Procedure

Groups 1–4: Examine a drop of the algal culture provided on a slide in the microscope (at low magnification without using a coverslip) to see if the flagellates are actively swimming. Take 1 mL of culture in a test tube and add 0.5 mL of Lugol's Iodine. Then, using a Pasteur pipette, add 1 drop of antipyrine and then 1 drop of $FeCl_2$. Allow the cells to settle for a few minutes in the tube. From your fixation, transfer a drop to a slide and affix a coverslip. Examine the flagella using high magnification. Count the number of flagellated cells from 100 individuals. Using a graduated slide, calculate the real length represented by the ocular divisions of the calibration ocular provided. Now measure the length of 50 flagella from 50 separate cells. Try to choose straight flagella. Fill the homogenizer to the mark with algal culture and set the cylinder in a plastic cup of ice water. Apply 20 strokes of the homogenizer piston and then examine the algae. Be careful when using the homogenizer, as a vacuum can be created on the outward stroke, suddenly pulling the plunger inwards and breaking an expensive device. If the cells are still motile, repeat the homogenizer treatment until at least 90% of the cells are nonmotile. Make a fixation as described above and quantify the percentage of deflagellated cells. You are aiming to optimize the rate of deflagellation without damaging the cells by excessive homogenizer treatment. The optimal procedure is dependent on the type and size of the algae and also on the condition and make of the homogenizer. When you are satisfied with the amputation of the flagella, proceed as outlined for each group below.

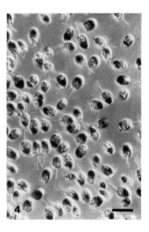

Fig. 32.4. Quadriflagellate cells of *Tetraselmis striata* prior to commencement of the experiment (scale bar = 20 μm).

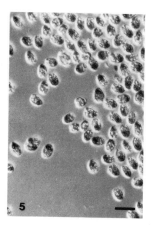

Fig. 32.5. Deflagellated cells immediately after treatment with the homogenizer (scale bar = 20 μm).

Group 1: Install the deflagellated algae in a test tube in the original culture cabinet or incubator at 20°C. After 15 min mix the algal suspension well and remove a 1 mL subsample. Prepare a fixation as previously described and examine the flagella. Repeat this procedure every 15 min (if using *Tetraselmis* then every 30 min), measuring the length of at least 30 flagella from 30 cells. It may be useful to seal the slides with nail varnish so that they do not dry out. Plot flagellar length versus time. From the last fixation, calculate the percentage of cells that regenerated flagella.

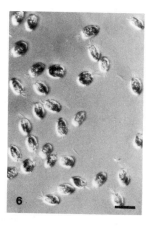

Fig. 32.6. Cells with partially regenerated flagella 90 min after deflagellation (scale bar = 20 μm).

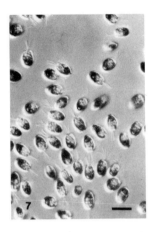

Fig. 32.7. Cells with almost fully regenerated flagella 180 min (incubation temperature = 20°C) after deflagellation (scale bar = 20 μm).

Group 2: Centrifuge the deflagellated cells at ca. 500 × g for 5 min. Decant the supernatant and gently resuspend the algal pellet in the cycloheximide medium (CAUTION –. Wear disposable gloves when handling cycloheximide as it is toxic). Examine the living cells in the microscope. Return the cells to the 20°C incubator. Make measurements of the flagellar length every 15 min. After 1 h, centrifuge the cells again, decant the cycloheximide solution and resuspend the algae in fresh media two times. Continue making flagellar length measurements every 15 min. The results of Group 1 can be used as your control data.

Group 3: Incubate your deflagellated cells at 15°C and make flagellar length measurements every 15 min the same as described for Group 1. Use the results of Group 1 as your control data.

Group 4: Before returning your deflagellated algae to the incubator, wrap the tube carefully in aluminum foil to prevent illumination throughout the experiment. Measure the flagellar length at 15-min intervals as described for Group 1. Use the results of Group 1 as your control data.

Questions

1. Is the rate of regeneration linear, and, if not, what could you surmise from the observed kinetics?
2. What is the effect of the protein synthesis inhibitor on regeneration? How does this relate to your observations of normal regeneration?

3. What influence has temperature on the regeneration rate and how do you explain this?
4. Is flagellar regeneration dependent on illumination? What might you infer from your observations about the synthesis of flagellar precursor substances?

REFERENCES

Amos, W. B. & U. G. Duckett (1982). *Prokaryotic and Eukaryotic Flagella.* Cambridge University Press, London.

Bouck, G. B. & S. J. Chen (1984). Synthesis and assembly of the flagella surface. *J. Protozool.* 31, 21–24.

Gibbons, I. R. (1983). Introduction to dynein workshop. *J. Submicrosc. Cytol.* 15, 177–180.

Gibbons, I. R. & A. J. Rowe (1965). Dynein: a protein with adenosine triphosphatase activity from cilia. *Science,* 149, 424–426.

McFadden, G. I. & R. Wetherbee (1985). Flagellar regeneration and associated scale deposition in *Pyramimonas gelidicola* (Prasinophyceae, Chlorophyta). *Protoplasma* 128, 31–37.

McFadden, G. I. & M. Melkonian (1986). Golgi apparatus activity and membrane flow during scale biogenesis in the green flagellate *Scherffelia dubia* (Prasinophyceae) I: Flagellar regeneration. *Protoplasma* 130, 186–198.

Melkonian, M. & H. Robenek (1984). The eyespot apparatus of flagellated green algae: a critical review. *Progress in Phycological Research* 3, 193–268.

Moestrup, Ø. (1982). Flagellar structure in algae. A review with new observations in the Chrysophyceae, Phaeophyceae, (Fucophyceae), Euglenophyceae, and in *Reckertia. Phycologia* 21, 427–528.

Remillard, S. P. & G. B. Witman (1982). Synthesis, transport and utilization of specific proteins during flagella regeneration of *Chlamydomonas. J. Cell Biol.* 93, 615–631.

Ringo, D. L. (1967). Flagellar motion and fine structure of the flagellar apparatus in *Chlamydomonas. J. Cell Biol.* 33, 543–571.

Rosenbaum, J. L. & F. M. Child (1967). Flagellar regeneration in protozoan flagellates. *J. Cell Biol.* 34, 345–364.

Rosenbaum, J. L., J. E. Moulder & D. L. Ringo (1969). Flagellar shortening and elongation in *Chlamydomonas. J. Cell Biol.* 41, 600–619.

Satir, P. (1974). How cilia move. *Scientific American* 231(4), 44–53.

Satir, P. (1984). The generation of ciliary motion. *J. Protozool.* 31, 8–12.

Satir, P. & G. K. Ojakian. (1979). Plant cilia. In, W. Haupt, & M. E. Feinleib (eds.). *Physiology of Movement* (Enc. Plant Physiol. 7), Springer, Berlin, pp. 224–249.

Stebbings, H. H. & J. S. Hyams (1979). *Cell Motility,* Longman, London.

Witman, G. B. (1975). The sites of *in vivo* assembly of flagellar microtubules. *Proc. N.Y. Acad. Sci.* 253, 178–191.

NOTES FOR INSTRUCTORS
Materials

Algae and media. The experiment requires ca. 50 mL of healthy culture (cell density ca. 10^6 mL^{-1}) per group. The flagellates should be motile prior to the lab session. This will be best achieved by keeping the culture in an active state of growth and timing the light:dark cycle (14:10 h) so that the light period begins 2–3 h before the class starts. The culture should be incubated at at least 20°C to achieve complete regeneration within a session. Two species of algae are recommended: *Chlamydomonas reinhardtii* (e.g., UTEX Culture Collection strain # 80) grown in Medium 1 (Sager and Granick, "Nutritional Studies in *Chlamydomonas*," *Proc. N.Y. Acad. Sci* 56 (1953), 831–38); or *Tetraselmis striata* (Plymouth Culture Collection # 443) grown in ASP-2, Erdschrieber or f/2 media (McLachlan, J., *Phycological Methods,* ed. J. Stein [Cambridge University Press, London, 1973], 25–51). From our observations there is reason to believe that this experiment will work on most green flagellates; instructors could easily pre-test other algae in their collection if the suggested material is unavailable. The main proviso is that the majority of cells must be motile and actively growing to achieve synchronous regeneration.

Solutions. The fixation/flagellar staining requires a solution of Lugol's Iodine (dissolve 3 g of KI and 3 g of I_2 in 100 mL of distilled water); 4 mM antipyrine (crystals may form when used in seawater media so dilute to ca. 1 mM when using *Tetraselmis*); and 50 μM FeCl$_2$ (these solutions can be stored for several months).

Cycloheximide is obtainable from Sigma in 1 g vials and a fresh stock solution should be made up in distilled water prior to the class and then diluted with sterile medium to 1 μg mL^{-1} *(Chlamydomonas)* or 0.1 μg mL^{-1} *(Tetraselmis).* This chemical is toxic and standard laboratory precautions should be observed in the handling. A small amount of fresh medium should be on hand to resuspend the cells after the cycloheximide treatment.

Equipment.

1. homogenizer: Kontes – Scientific Glassware/Instruments, Vineland, New Jersey. "Tissue grind tube" K885512-0022. This apparatus is known as a tissue homogenizer and comprises a glass cylinder like a test tube (the sizes range considerably) and a close fitting plunger of either glass or teflon. We find that various models will work with a

range of algae. Larger homogenizers (20–30 mL) will allow each group to deflagellate enough culture for their experiment in one run. If you have a lot of students and few homogenizers and are doing both *Chlamydomonas* and *Tetraselmis,* start the *Tetraselmis* first. The "fit" of the plunger affects the shear force created so it is best to do a trial run with the homogenizers available. Homogenizers are expensive and best borrowed, but care should be taken during use, particularly on the outward stroke as cavitation can pull the plunger back unexpectedly.

2. microscope(s): 1 per group with 100 × oil immersion and, if available, phase contrast.
3. oculars: 1 per group. Adjustable types are optimal.
4. calibration slide: At least one required and it may be expedient for the instructor to calibrate the oculars beforehand.
5. sundries: Slides, coverslips, immersion oil, Pasteur pipettes, teats, large test tubes, racks, graph paper, nail varnish.
6. centrifuge: small bench-top models or even hand-operated centrifuges generating about 500 to 1,000 × g are adequate to sediment the algae used (only needed if doing cycloheximide inhibition).

Scheduling

This experiment can be done within a 3-h laboratory session if well managed. *Tetraselmis* takes almost 3 h to regenerate completely, so students using this organism must work rapidly in the first stages.

AUTHOR INDEX

(Numbers in italics indicate figures.)

(Numbers in italics indicate figures.)

TAXONOMIC INDEX

(Numbers in italics indicate figures.) noted on gg.